Textile Progress

2009 Vol 41 No 4

Pulse-jet filtration: An effective way to control industrial pollution Part I: Theory, selection and design of pulse-jet filter

Arunangshu Mukhopadhyay

The Textile Institute

SUBSCRIPTION INFORMATION

Textile Progress (USPS Permit Number pending), Print ISSN 0040-5167, Online ISSN 1754-2278, Volume 42, 2010.

Textile Progress (www.tandf.co.uk/journals/TTPR) is a peer-reviewed journal published quarterly in March, June, September and December by Taylor & Francis, 4 Park Square, Milton Park, Abingdon, Oxon, OX14 4RN, UK on behalf of The Textile Institute.

Institutional Subscription Rate (print and online): $335/£176/€267
Institutional Subscription Rate (online-only): $319/£168/€254 (plus tax where applicable)
Personal Subscription Rate (print only): $123/£63/€98

Taylor & Francis has a flexible approach to subscriptions enabling us to match individual libraries' requirements. This journal is available via a traditional institutional subscription (either print with free online access, or online-only at a discount) or as part of the Engineering, Computing and Technology subject package or S&T full text package. For more information on our sales packages please visit www.tandf.co.uk/journals/pdf/salesmodelp.pdf.

All current institutional subscriptions include online access for any number of concurrent users across a local area network to the currently available backfile and articles posted online ahead of publication.

Subscriptions purchased at the personal rate are strictly for personal, non-commercial use only. The reselling of personal subscriptions is prohibited. Personal subscriptions must be purchased with a personal cheque or credit card. Proof of personal status may be requested.

Ordering Information: Please contact your local Customer Service Department to take out a subscription to the Journal: **India**: Universal Subscription Agency Pvt. Ltd, 101–102 Community Centre, Malviya Nagar Extn, Post Bag No. 8, Saket, New Delhi 110017. **USA, Canada and Mexico**: Taylor & Francis, 325 Chestnut Street, 8th Floor, Philadelphia, PA 19106, USA. Tel: +1 800 354 1420 or +1 215 625 8900; fax: +1 215 625 8914, email: customerservice@taylorandfrancis.com. **UK and all other territories**: T&F Customer Services, Informa Plc., Sheepen Place, Colchester, Essex, CO3 3LP, UK. Tel: +44 (0)20 7017 5544; fax: +44 (0)20 7017 5198, email: tf.enquiries@tfinforma.com.

Dollar rates apply to all subscribers outside Europe. Euro rates apply to all subscribers in Europe, except the UK and the Republic of Ireland where the pound sterling price applies. If you are unsure which rate applies to you please contact Customer Services in the UK. All subscriptions are payable in advance and all rates include postage. Journals are sent by air to the USA, Canada, Mexico, India, Japan and Australasia. Subscriptions are entered on an annual basis, i.e. January to December. Payment may be made by sterling cheque, dollar cheque, euro cheque, international money order, National Giro or credit cards (Amex, Visa and Mastercard).

Back Issues: Taylor & Francis retains a three year back issue stock of journals. Older volumes are held by our official stockists to whom all orders and enquiries should be addressed:
Periodicals Service Company, 11 Main Street, Germantown, NY 12526, USA. Tel: +1 518 537 4700; fax: +1 518 537 5899; email: psc@periodicals.com.

The 2010 US Institutional subscription price is $335. Periodical postage paid at Jamaica, NY and additional mailing offices. US Postmaster: Send address changes to TTPR, C/O Odyssey Press, Inc. PO Box 7307, Gonic NH 03839, Address Service Requested.

Subscription records are maintained at Taylor & Francis Group, 4 Park Square, Milton Park, Abingdon, OX14 4RN, United Kingdom.

For more information on Taylor & Francis' journal publishing programme, please visit our website: www.tandf.co.uk/journals.

CONTENTS

Textile Progress
Vol. 41, No. 4, 2009, 195–315

Pulse-jet filtration: An effective way to control industrial pollution Part I: Theory, selection and design of pulse-jet filter

Arunangshu Mukhopadhyay*

National Institute of Technology, Jalandhar, India

(Received 22 October 2009; final version received 14 November 2009)

Pulse-jet filtration is described as one of the most efficient technologies in controlling industrial pollution across the world. The monograph provides the fundamental concept of design and development of pulse-jet filters under varied siutations. For successful running of a filter unit, a comprehensive knowledge base as regards a selection of design and development of filter media is essential; thus, this is incorporated in the monograph. I also discuss technical and commerically attractive solutions for successful operation of industries integrated with pollution control equipment maintaining clean air requirements.

Keywords: filter media design; gaseous emission; industrial pollution; particulate emission; pulse-jet filter

1. Industrial air pollution and role of filter unit

1.1. Industrial emission

The environmental issue has become a major subject in the last few decades, affecting science and technology of the entire world due to serious environmental impacts caused by air pollution. Environmental pollution has negative influences on human health, on ecological systems, the greenhouse effect, the ozone layer, etc. One of the major causes of pollution is industrialization; and due to the technological and economic limitations, the atmosphere is used as a sink for the disposal of airborne waste. Inevitably, in many industrial processes, the separation of solids from gases by filter unit is an essential part, which contributes toward the purity of the product, energy saving, improvements in the process efficiency, recovery of precious material, and, most importantly, the controlling of air pollution. Many of the problems linked to the emission of particulate matter into the atmosphere have resulted in legislation, which is becoming more and more rigorous. There is also the increasing importance of legislation requiring operators to provide information about gaseous and particulate substances, which are released into the environment from the processes operators' control. The concerns which lie behind the legislation relate to the following [1]:

- Health effects on human beings and animals;
- Atmospheric effects (visibility and turbidity affecting transport safety, property values, and aesthetics; thermal air pollution; atmospheric deposition (i.e. acidic, nitrogen, mercury (Hg) deposition etc.); stratospheric ozone depletion; global warming; etc.);
- Welfare effects (agricultural crops, ornamental plants, and trees); ecological effects; effects on materials, such as building decay.

*Email: arunangshu@nitj.ac.in

ISSN 0040-5167 print/ISSN 1754-2278 online

DOI: 10.1080/00405160903437948
http://www.informaworld.com

There are good resource materials available that deal with the science of atmospheric chemistry, effects on health and the environment, and the regularity practices employed in achieving air quality goals [1–3]. In commensurate with legislation, the regulatory framework encompasses three general categories: (1) environmental quality standards, setting acceptable levels of pollutants in the environment; (2) performance standards, limiting discharge of specific substances in the environment; and (3) design standards, prescribing the control methods to be used. Ways of eliminating particulate material from gases and controlling gaseous emissions are therefore of great industrial importance. Particulate matter is one of the primary air pollutants and carries some of the most toxic materials hazardous to human health. They are produced as unwanted byproducts in various industrial processes utilizing fossil fuels, such as boilers, steel and cement manufacturing processes, incinerators and combustors, to name just a few [4,5]. The emissions of particulate matter are variable, with particle concentration ranging from less than 1 g/m^3 to more than 250 g/m^3, and the particles size are predominantly very fine (0.1–25 μm). As most fine particles occurring in industrial processes are hard to collect with conventional technologies, they are discharged into the atmosphere. Consequently, they threaten the health and lives of residents and employees of the areas where these industrial plants are located. Particles less than 2.5 μm in diameter ($PM_{2.5}$) are referred to as fine particles and are believed to pose the greatest health risks [6–8]. Because of their small size (approximately 1/30th the average width of a human hair), fine particles can lodge deeply in the lungs. Health studies have shown a significant association between exposure to fine particles and premature death. Fine particles with size smaller than 2.5 μm contain sulfates and nitrates, which considerably reduce the visibility in the atmosphere.

The assessment of air quality at the time of emission (discharge of pollutant in the form of solids, liquids, or gases into the atmosphere) is judged by comparing it with emission limits. Particulate emission is generally expressed in mg/Nm^3; it is also expressed otherwise as g/t or kg/t of production or ng/Joule of heat input. Particulates are the non-gaseous constituents of the gas stream being filtered; they can be liquid or solid and can originate from dispersion or condensation processes. Figure 1 shows the particle size range for different particulate pollutants. Emission limits vary depending on the industrial process. Table 1 shows emission limits for various industries in accordance with German

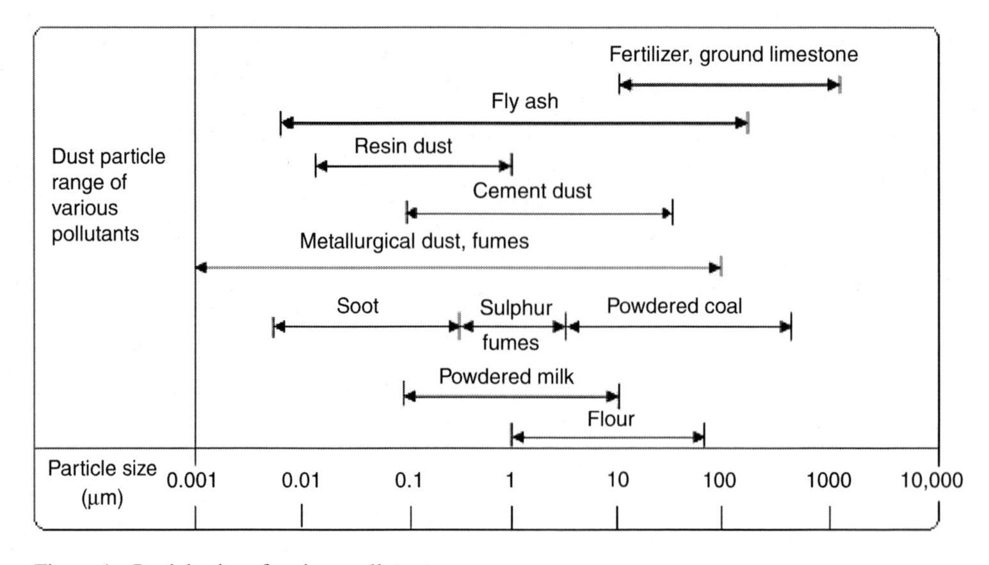

Figure 1. Particle size of various pollutants.

Table 1. Emission limits for various industries – German standards.

Type of industry	Particulate (mg/Nm^3)
Furnace for coal, coke, coal briquette, pleat, wood, wood processing residue (<50 MW)	50/150
Combustion with solid fuel at large firing installation ($\geq$50 MW and $\geq$ 100 MW)	50
Combustion with liquid fuel at large firing installation ($\geq$50 MW and $\geq$ 100 MW)	50
Combustion with gaseous fuel at large firing installation ($\geq$50 MW and $\geq$ 100 MW)	5
Briquetting brown and hard coal	75
Dry distillation of hard coals	25/20
Production of lead accumulator	0.5
Production of aluminium	30
Furnace soots	20
Burning bauxite, dolomite, gypsum limestone, diatomite magnesite, fine clay	10
Production of nonferrous unfired metals	20/10
Aluminium smelter	20
Smelters including refining nonferrous metals and their alloys other than aluminium	20/10
Grinding or mechanically mixing, packing, or refilling plant protective pesticides or their active ingredients	5
Ferro alloys in electrothermal or metallothermal processes	20
Production of steel in converter, electric arc furnaces, and vacuum melting systems, smelting steel, or cast iron	20/50
Foundries and nonferrous metal	20

Standards (in the early 1980s). It may be added that both Germany and the United States have extensive emission standards for different industries. However, Germany is the first and perhaps the only country so far to promulgate the act relating to closed substance cyclic waste management and environmentally compatible disposal of waste covering the areas of avoidance, recycling, and waste disposal.

Standards for emission limits are different in different countries (Table 2) [9]. The comparative data presented in Table 2 is based on the indicated references only, which may not cover the complete norms of the countries covered.

Looking back into the history of enhancement of air pollution control regulations one finds that countries world over were not particularly careful about them till the late 1950s [9]. However, during the last 50–60 years countries in Europe and North America, and many other developed countries like Germany have made significant efforts to introduce and implement exhaustive standards in air pollution control. The basic purpose of implementing the industrial air pollution control legislation is to minimize the discharge of particular substance into the environment. Regulation is the rule or order having the force of law issued by the authority or the government. The objectives of the regulating authority is to see that standards are being met by industries and if not, to take suitable measures for their enforcement.

The air quality directive of the European Community (1999/30/EC) [9] stipulates an average annual limit of 40 $\mu g/m^3$ from 2005 on and sets an indicative limit of 20 $\mu g/m^3$ for the year 2010. A Directive Proposal (2005/0183 CODE) [10], which is being revised, sets the average annual limits for PM_{10} (PM_{10} is the mass concentration of particles with aerodynamic diameters smaller than 10 μm) and $PM_{2.5}$ at 14 and 10 $\mu g/m^3$, respectively. The United States, after the process of update and review carried out by the Environmental Protection Agency (EPA) at the end of 2004, had set a limit on the annual mean for PM_{10} and $PM_{2.5}$ at 50 $\mu g/m^3$ and 15 $\mu g/m^3$, respectively [11,12]. This indicates that the legislation

Table 2. Atmospheric emission limits of particulate pollutants from stationary sources for different countries.

Pollutants	Australia	Austria	Brazil	Denmark	France	Germany	India	Mexico	United Kingdom	Poland (g/GJ)*	Singapore	Switzerland (mg/Nm^3)	United States (ng/J)*
Particulate (mg/Nm^3)	50	12–50	100	5–200	12–150	0.5–150	150–400	250	50–100	70–1370	200	10–150	13–130

*Emission is expressed in terms of heat input.

concerning emissions is ever stricter. However, the number of concentration, and not the mass concentration, is believed to affect human health.

Apart from the particulate matter, the system should entail to removal/control of harmful gases and obnoxious odors wherever applicable. In the emission, there could be various inorganic and organic gases, Hg, and trace metals. In the case of gaseous pollutants, the most important of those discharged in a combustion process are carbon monoxide (CO), carbon dioxide (CO_2), sulfur dioxide (SO_2), and nitrogen oxide (N_XO_Y). The particulate matter (fly ash) and SO_2 are the major industrial pollutants. Emission limits for various industrial air pollutants can be obtained from elsewhere [13]. Since the passage of the Clean Air Act in 1970, emissions of air pollutants, such as fly ash, SO_2, NO_x, and Hg emissions from the coal-burning electric power industry, have decreased significantly. Since fossil fuel burning for power generation produces about 90% of man-made emission of SO_2 and (N_xO_y), this causes environmental degradation in the form of acid rain, which has been a major concern for controlling the emission.

Apart from the inorganic gases, there are a large number of industrial processes where large quantities of organic vapors are emitted into the atmosphere. Volatile organic compounds (VOCs) are gases or vapors emitted by various solids or liquids, and can be generated from industrial processes, many of which have short- and long-term adverse health effects. From a chemistry viewpoint, *volatile organic compound* means any organic compound (all chemical compounds containing carbon with exceptions) that is volatile (evaporating or vaporizing readily under normal conditions). VOCs have low water solubility, and high enough vapor pressures under normal conditions to significantly vaporize and enter the atmosphere. A wide range of carbon-based molecules, such as aldehydes, ketones, and other light hydrocarbons are VOCs. The compounds such as trichloroethylene (TCE, C_2HCl_3), toluene ($C_6H_5CH_3$), benzene (C_6H_6), etc. are known as environmental pollutants attracting social attention. Some VOCs are directly harmful (are carcinogenetic, cause discomfort, nervous paralysis, etc.) and cause social destructions such as ozone hole destruction, green house effect, and so on. The conventional technologies for eliminating these pollutants are carbon adsorption, catalytic oxidation, and thermal incineration [14]. To remove these toxic gasses, many technologies have been developed, but they are not very successful and do not satisfy strict social norms of the day. Further, these technologies have cost and maintenance problems.

Recently, gaseous Hg, which is a representative hazardous air pollutant (HAP) among the air pollutants produced from the combustion of flue gas, has prompted widespread concern throughout the world. Mercury is one of the most toxic pollutants existing in the ecosystem. Approximately 25% of the global municipal waste production is processed in incinerators. During incineration, practically all of the mercury is evaporated and must be removed in the flue gas treatment section. The Hg concentration in municipal/industrial waste and the consequences of its presence in the incineration have decreased considerably during the last decade. Presently the Hg concentration in municipal solid waste is approximately 2 mg kg^{-1} and a number of processes guarantee Hg concentration of less than 50 μg Nm^{-3} in clean gas, the present limit of Hg emission [15]. Gaseous Hg is basically classified into elemental mercury (Hg^0) and oxidized mercury (Hg^+, Hg^{2+}). Unlike oxidized Hg, the difficulties in the removal caused by its insolubility in water and poor reactivity with other species have made the Hg^0 a major target of research in the field of gaseous Hg removal.

The design and development of filtration equipment should be more cost-effective when operating at a higher performance level. The filter unit should remove solid and liquid particulates with sizing well below 1 μm; there is now an additional requirement for collecting vapor phase materials to meet the latest regulatory emission levels [16]. The selection of gas-cleaning devices is influenced by many parameters such as efficiency, characteristics of particulate and gaseous matter, capital investment, availability of space, power, and water, operating and maintenance charges, construction complexity, and estimated life.

1.2. Impact of fine particulate matter

Airborne dust particles are rarely homogeneous and vary greatly in size and shape, and their chemical composition is determined by factors specific to the source and location of emissions. The combined effects and interactions of various substances mixed with particulates have not been established (except for SO_2), but they are believed to be significant, especially when long-term exposure takes place [1]. Overall, the risk of adverse health effects from inhaled particles is a function of size and concentration of fine particulate matter in the air, the duration of exposure, the penetration and deposition of particles in the regions of the respiratory tract, and the body's biological responses to the deposited materials. These factors relate to the amount of pollutant that actually enters the body over a specified period.

In general, air pollution could lead to lung cancer, asthma, lung infections, chronic obstructive pulmonary disease (COPD), coronary artery disease, heart failure, heart-rhythm problems, and various other diseases. The respiratory system is particularly sensitive to air pollutants because much of it is made up of exposed membrane. Lungs are anatomically structured to bring large quantities of air (on average, 400 million liters in a lifetime) into intimate contact with the blood system to facilitate the delivery of oxygen (O_2). Although a lung is protected by the defense mechanism in respiratory system (Table 3), it is difficult to control the entry of finer particles and the percentage of their retention is high. For the particle size below 10 μm, lungs are prevented with the help of cilia (tiny hair-like structures), mucus, and macrophages, which act to remove harmful substances before being deposited inside the lungs. Approximately 40% of the particles between 1 μm and 2 μm in size are retained in the bronchioles and alveoli (Figure 2). Particles ranging in size from 0.25 to 1 μm show low rates of retention in lungs because many such particles are breathed in and out during respiration (Figure 3). However, particles below 0.25 μm in size show increase in retention because of Brownian motion, which results in impingement [1,12,17,18]. Despite the defense mechanism in the respiratory track, many fine particles enter pulmonary tissues. Within the alveoli and bronchioles of the pulmonary region, specialized cells called phagocytes and macrophages ingest deposited matter. Although the mechanism affects the clearance from the respiratory system, they may expose other body systems to toxic materials. They may also result in tissue damage when phagocytes are destroyed and spill their contents when attempting to ingest toxic particles. Contrary to the

Table 3. Particle size and body-defense mechanism.

Particle size	Description	Mechanism
Particulate over 10 μm	Coarse dust	Hairs at the front of the nose remove all particles
2 to 10 μm	Fumes, dust, smoke particles	Movement of cilia sweeps mucus upward, carrying particles from windpipe to mouth, where they can be swallowed
Less than 2 μm	Aerosol, fumes	Phagocytes and macrophages (part with white blood cells) in the lung attack some submicron particle

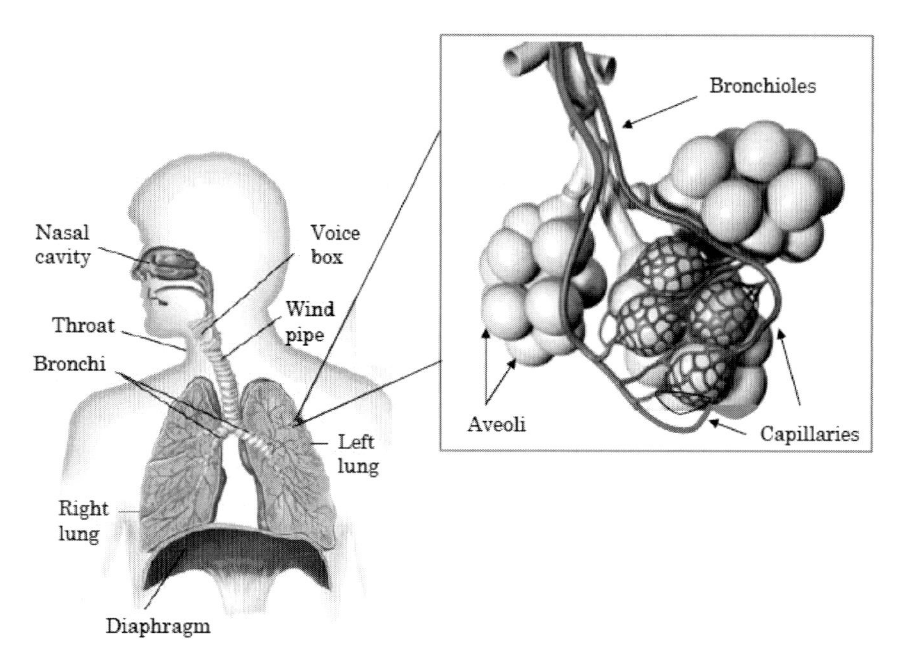

Figure 2. Human lungs and alveoli.

larger particles, nanoparticles are not assimilated by the microphages but can directly pass through the cell wall.

In the case of deposited particulates inside the lungs, there is movement of very small particles of matter, or molecules, across the semipermeable membranes of blood capillaries as the pressure inside the capillaries is much greater than on the other side of the membrane surrounded by fluid. This higher pressure forces particles through the capillary walls by means of hydrostatic pressure. The inhalation of air pollutants eventually leads to their absorption into the bloodstream and transportation to the heart affecting the cardiovascular system. The cardiovascular system has two major components: the heart and a network of blood vessels. The cardiovascular system supplies the tissues and cells of the body with nutrients, respiratory gases, hormones, and metabolites, and removes the waste products of cellular metabolism as well as foreign matter. It is also responsible for maintaining the optimal internal homeostasis of the body and the critical regulation of body temperature

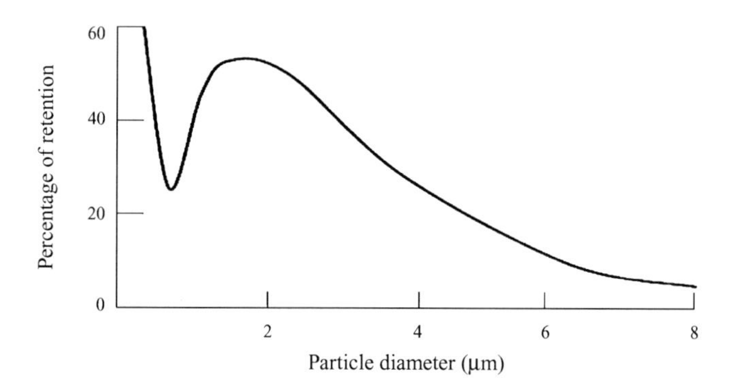

Figure 3. Retention of particulates in lungs [17].

and pH. A wide spectrum of chemical and biological substances may interact directly with the cardiovascular system to cause structural changes, such as degenerative necrosis and inflammatory reactions. Some pollutants may also directly cause functional alterations that affect the rhythmicity and contractility of the heart.

Air pollution can also damage the alveoli – the individual air sacs in the lungs where O_2 and CO_2 are exchanged. Airway tissues, which are rich in bioactivation enzymes, can transform organic pollutants into reactive metabolites and cause secondary lung injury. Lung tissue has an abundant blood supply that can carry toxic substances and their metabolites to distant organs. In response to toxic insult, lung cells also release a variety of potent chemical mediators that may critically affect the functioning of other organs such as the cardiovascular system. In a recent study, it has been observed that lungs inflamed by pollution secrete interleukin-6, an immune system compound that sparks inflammation and makes blood more likely to clot. This response may also impair lung function. Many chemical substances may cause the formation of reactive O_2. The production of O_2-free radicals in heart tissues has been associated with arrhythmias and heart cell death. There are various literatures [1,18] from which more details about the overall impact of particulates and gases on human health can be obtained.

1.3. Role of filter unit and allied development

The successful operation of an industry and integration of the pollution-control equipment, while still maintaining clean air requirements, has become a real challenge for the production sector. A classification of different aerosol filters is shown in Figure 4. In general, an aerosol is defined as the suspension of solid and liquid particles in a gaseous medium with negligible rate of fall. Basic constriction of various aerosol filters can only control/arrest particulate matter and have no effect on gaseous component. At the source of emission, different types of filters can be used depending upon the type of pollutants and the regulatory control needs. Initially, when the emission norms were not stringent, and technology of fabric filter and electrostatic precipitator was not in practice, mechanical dust collectors like cyclones and multi-cyclones were predominant, generally for smaller boilers. In general, mechanical separators possess low efficiency, particularly for small particles and, therefore, with the introduction of more and more stringent regulations, gradually fabric filters and electrostatic propitiator have become major particulate collectors.

A recent change in the design of milk powder plants has seen the traditional cyclone system used for fine recovery replaced by washable baghouse systems. Washable baghouses, with their ability to be cleaned-in-place (CIP), have changed the perception of baghouses as being a source of microbial contamination. Consequently, all fines collected by them retain the microbial product quality of the dryer chamber powder, thereby increasing the profits.

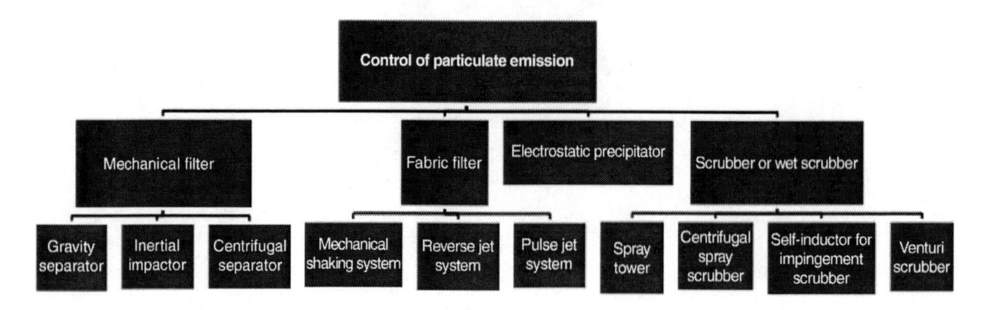

Figure 4. Classification of different aerosol filters.

Because all fines are collected in a single stage, fewer unit operations are required. This leads to reduced building space requirements, lower capital costs, and simplified plants [19]. The final advantage of the new system is that baghouse collection is a gentler form of product collection when compared with cyclones. This is more suited to the high fat and high protein powders of the dairy industry. However, presently, the mechanical type of collectors is used in the process line to remove the coarser particulate matter and therefore reduce the load on final filter unit.

Historically, ESPs were the first choice for large industrial boilers and utility boilers of all kinds because of the limitations of fabric filters until the advent of suitable nonwoven fabric for filtration. Before the 1950s, use of woven fabric filters, particularly in process industries where adverse conditions like high-temperature and aggressive chemicals are prevalent, was limited due to their unsuitability. Woven wool and cotton, which are not suitable for the above duty conditions were only used. During the days when suitable synthetic fibers were not developed for high-temperature applications, reverse air-cleaned a woven fabric filter made out of glass fiber was the only choice. With the development of nonwoven fabrics composed of polyphenylene sulphide (PPS – Ryton), Copolyimide (P84), Polyamide (Nomex), etc., use of pulse-jet filters gradually become more popular. The advent of synthetic man-made fiber beside the use of nonwoven fabric has brought about a revolution in fabric filter technology [3]. In contrast with woven fabric, nonwoven fabric collector does not require a dust cake to achieve high collection efficiency and it allows high permeability even at high levels of efficiency. However, at the beginning of the use of fabric filter, there was substantial bag failure attributed to abrasion, acid attack, poor manufacture, and improper tensioning, but the trend showed a continuous and significant decline since 1979. Today fabric filter has established its place over other types of filters for similar applications. Fabric filters enjoy the advantage of their independence from fuel and ash characteristics, as any variation in operating parameters, like gas flow and temperature (as long as it is within the limit of fabric used), do not cause any change in emission.

It is also important to control the gaseous emission. Wet electrostatic precipitator and venturi scrubber, though primarily used for dust collection, in the process also control gaseous pollutants. The control of inorganic and organic gaseous pollutants from stack gases depends on their properties. The methods of control include combustion, absorption, adsorption, closed collection and recovery systems, and masking and counter action (for odors) [20–22].

Many countries have been focusing on the development of improved process for the removal of SO_2 and N_xO_y, and are simultaneously tightening the pollution norms. In most of the cases, oxides of sulfur (SO_x) and nitrogen (N_xO_y) are handled separately, although there are instances where simultaneous removal of SO_x and N_xO_y has been successful [23,24]. Removal of SO_2 is commonly known as desulphurization ($DESO_x$) or flue gas desulphurization (FGD), while that of N_xO_y is called denitrification ($DENO_x$). SO_2 control is of three types, viz., pre-combustion, in-combustion, and post-combustion. Pre-combustion technology is the cleaning of coal (coal beneficiation), which can remove 10–30% of sulfur content in coal [5]. The most widely used SO_2 control is post-combustion technology where an absorbent (wet, dry, or semidry) is used. The denitrification as opposed to desulphurization is a relatively new activity. The first denitrification plant using the selective catalytic reduction (SCR) process was started in 1985 at 460 MW Altbach power station of Neckarwerke AG, Germany [23]. Post-combustion SCR technology has now become common around the world.

In the case of fabric filters, through specialized filters, such as catalytic filters, or through absorption by specifically developed filter cakes, some forms of gaseous emission can be controlled. The combined use of fabric filters with sorbent injection systems has been utilized for many years in the municipal incinerator and other industries as a way to enhance the removal of Hg and other pollutants like dioxins, furans (VOCs), and a wide range of heavy metals [25]. Fine particle and trace element emissions from energy production are associated with significant adverse human health effects. The baghouse collection efficiencies of particles, and trace elements (As, Hg, Se, Cd, chromium (Cr), Cu, aluminium (Al), vanadium (V), zinc (Zn), Mn, Fe) are being studied [26]. There are various patent technologies for the removal of targeted gaseous and trace metals pollutants from the industrial processes [27–32]. Pollution-control strategies to reduce Hg emission can be highly interrelated. Strategies to reduce emissions of any one pollutant from power generation can have effects on emissions of other pollutants. The cost and other impacts of control strategies for these pollutants are also highly interdependent. Different technologies encompass three basic aspects: pre-combustion, post-combustion, and energy efficiency [33].

Over the past many years, the industrial filtration process has seen tremendous improvements. Factors which have influenced this evolution are as follows [34–36]:

- New stricter emission regulations for fine dust particulate;
- Increased filtration capacity within the same or smaller area;
- Increased use of dust collectors for product collection with simultaneous control of gaseous pollution.

In the recent past, stress is also being given to having control over the source to reduce or eliminate waste. Efforts are also required to reshape the technological approach to pollution control and prevention. Prevention is frequently more cost-effective than control. In many cases, a control system alone is not able to cope with the situation; prevention has also become a major aspect of research. In practice, waste removal does not signify elimination of treatment and a system without disposal – a potential environmental threat. Therefore, it is also very important to focus attention on the following allied areas:

- Change in process philosophy;
- Change in plant operation;
- Introduction of recycling;
- Raw material substitution;
- Modifying the end product.

Measures such as improved process design, operation, housekeeping, and other management practices can reduce emission. One must also keep oneself updated with developments in the process technology because characteristics of gas and dust change with the change of technology of manufacturing processes. By improving combustion efficiency, the extent of emitted particulate matter can be significantly reduced. Proper fuel-firing practices and combustion zone configuration, along with adequate amounts of excess air, can achieve lower emission. The Polish coal fired in the pulverized coal-fired boiler for combined heat and power production results in an unusually low fraction of fine particles compared with other coals [37]. However, advanced coal combustion technologies such as coal gasification and fluidized bed combustion are the examples of cleaner processes that may lower the product of incomplete combustion by approximately 10%. Dust is finer and resistivity is higher in the case of pulverized coal-fired boilers compared to fluidized bed combustion boilers. In addition, enclosed coal crushers and grinders emit lower particulate matter [13,38]. Yet another approach in combustion engine [39] has been proved to be effective. By modifying the coal particles' size in a pulverizer and changing the combustion process, fly ash emission can be controlled to become coarser, limiting the finer particles, which are more harmful to health and also difficult to collect.

Use of gaseous fuel (LNG) or oil-based processes also emit significantly fewer particles than coal-fired combustion processes. Low ash fossil fuels contain less noncombustible, ash-forming mineral matter and thus generate lower levels of particulate emissions. Lighter distillate, oil-based combustion results in lower levels of particulate emission than heavier residual oil. The choice of fuel is usually influenced by economic as well as environmental considerations [38]. A study [40] reveals that the conditioning of the ash with ammonia does not alter its physicochemical-mineralogical properties; although, from the scanning electron microscope (SEM) micrographs of the conditioned fly ash samples, the agglomeration of ash particles has been observed. Therefore, it is quite expected that filtration performance might improve through gas conditioning.

When considering change in process technology, basic research in combustion engineering shows that it is possible to design a combustion process with reduced N_xO_y emission [41,42]. Coal beneficiation in power plants has a substantial effect on SO_2 control [43]. Through coal washing in a thermal power plant, a greater degree of uniformity in particle size can be achieved in addition to reduction in ash content to an acceptable level. Furthermore, there is reduction in sulfur content of coal by removing pyritic sulfur, which takes the form of discrete particles. Organic sulfur containing 30–70% of total sulfur can only be eliminated by chemical processing. An alternative to coal cleaning [38] is the cofiring of coal with higher and lower ash content, which also results in reduction in particulate emission.

The average emission factor was found to be much higher in Chinese plants than in US plants. This is due to the burning of bituminous coal in Chinese plants, which has high Hg content and also because of low Hg removal efficiency of air pollution control devices of power plants [44]. The proportion of Hg in coal feedstock that is emitted by stack gases of utility power stations is a complex function of coal chemistry and properties, combustion conditions, and the positioning and type of air pollution control devices being employed. Key variables affecting performance of Hg control include chlorine and sulfur contents of the coal, the positioning (hot-side vs. cold-side) of the system, and the amount of unburned carbon in coal ash. Knowledge of coal quality parameters and their effect on the performance of air pollution control devices allow the optimization of Hg-capture benefits [45]. Mercury in bituminous coal is found primarily within iron sulfides, whereas lower-rank coal tends to have a greater proportion of organic-bound Hg. However, the mode of occurrence of Hg in coal does not directly affect the speciation of Hg in the combustion flue gas. Apart from the presence of chlorine and sulfur, the combustion characteristics of the coal influence the species of Hg that are formed in the flue gas and enter air pollution control devices [45,46]. Wang et al. [46] reported Hg speciation and emissions from five coal-fired power stations in China. Through analysis, it is found that the Hg speciation varied greatly when flue gas passed through the combination of fabric filter and electrostatic precipitator.

It is also important to reduce harmful waste by changing the reaction parameters like temperature, pressure, etc. in chemical processes. Also, recycling of waste material, which is hitherto disposed of, offers a new vista for more meaningful handling and use. There are also possibilities that the waste materials could be used in some other applications. For example, fly ash, due to its pozzolonic characteristics is used in cement making. Attempts are made to manufacture bricks from fly ash. Furthermore, fly ash due to it alkaline properties may be utilized to absorb SO_2 in place of lime/limestone. However, in all cases, large-scale usage is yet to come forth. Today technologists must think of an integrated environmental control approach where control devices have compatibility with the main machine, keeping in view prevention of pollution, effective control of pollutants, and waste disposal management [13,47].

2. Present status of pulse-jet filter in industry

2.1. General

For controlling industrial pollutants, typically in the range of 0.1–25 μm or higher, they have to be collected by several techniques [13]. As has been mentioned earlier, mechanical types of filter are, in general, effective for the removal of coarser particulate matter; these can be used to reduce burden of the final filter unit. For collecting small particulate matter, electrostatic precipitator, wet scrubber, and fabric filters are the only options. Among all the filters, the most efficient and versatile is the fabric collector, especially when processing very fine particles, which are slow to settle and by virtue of their greater light scatter, more visible to the naked eye. The overall collection efficiency of the existing devices is high for fabric filters, followed by electrostatic precipitators and wet scrubbers.

In the case of electrostatic precipitators (ESP), electrostatic devices used for cleaning gases exploit electrical forces to facilitate the removal of particulate matter. Electrical energy is required for gas ionization, particle charging, particle coagulation or agglomeration, or vapor condensation. The two most important mechanisms for charging particles by ionic current consist of the following [48,49]:

1. *Field charging*, whereby ions are driven to the particle due to electrostatic force caused by an external electric field. This force is balanced by the repulsive force of the charge imparted to the particle.
2. *Diffusion charging* is due to the kinetic energy of gaseous ions, which bombard the particle independent of the electric field.

A typical ESP incorporates two zones:

1. The charging zone, where the dust or aerosol particles are charged.
2. The collecting zone, where the charged particles are separated and transferred from the gas stream to a collecting electrode with subsequent removal into the collecting or receiving hoppers.

The arrangement of these zones led to two typical ESP design concepts: (a) a single-stage *conventional* ESP, where both zones are combined in a common area, and (b) so-called two-stage *design* ESP, where these zones are separated. In contrast with ESP, collection of particles by the fabric filter is purely governed by physical characteristics of constituent fiber and pore geometry defined by fibrous assembly.

The properties of a fabric filter that distinguish it from other types of filters include the following:

- The constituent fibers are randomly oriented and held semirigidly in the structure. Fiber characteristics, viz., fiber fineness, length, and cross-sectional area, its orientation and consolidation influence the filtration performance.
- Pore sizes and shape are determined by fiber arrangement and its consolidation. The size of pores is usually large compared to the size of the particles filtered unlike sieves or some membrane filter.
- An industrial filter operates under the mode of surface filtration; retaining of particles inside the structure by depth filtration deters the performance of the filter.
- The relatively low values of face velocity are typical for fabric filters.

- The capability of *in situ* regeneration, i.e. a fabric filter, can be repeatedly cleaned without removal by a technique designed into its mounting and housing.

In another important technology, scrubbers, wet, dry, or semidry, employ liquid or solid, or both in the form of fine droplets of liquid or solid to reduce particulates or gaseous pollutants, or both from the flue gases. In liquid scrubbers, gas and liquid are made to come into intimate contact so that dust or gas, or both can be collected or absorbed. Gas and liquid may be concurrent or countercurrent depending on the type of scrubber. Probably, scrubber is the only pollution control device that can effectively and economically remove dust and gas at the same time [13].

A comparison between filters for capturing small particles can be seen in Table 4. For better apprehension about the fabric filter, Table 4 compares the various control equipment, viz., fabric filter, electrostatic precipitator, and wet scrubber.

In selecting air pollution control equipment, both technical and economical considerations should be made. The selection should primarily concentrate on technical merits. As the technical selection is over, it is essential that economic factors (capital and operating costs) should play an important role. The specific reasons for wider acceptability of fabric filter can be summarized as follows [23,44,50–56]:

1. Fabric filters possess extremely high collection efficiency on both coarse and fine particulates. Fabric filters can be designed to collect particles in the sub-micrometer range with 99.9% control efficiency. Removal of very small particles ($PM_{2.5}$) at a very high level of efficiency is becoming increasingly important as more stringent emission controls are required.
2. The fabric filter is quite versatile as it can handle large varieties of dust differing in physical and chemical properties. It can capture all particles, not only those that can be charged electrically (as in ESPs). Performance of fabric filters is effective compared to electrostatic precipitators when the electrical resistivity of dust particles is very high. When burning the low sulfur coal, the fabric filters are frequently lower in capital and annual costs than electrostatic precipitators.
3. ESP is sensitive to change in operating parameters, such as temperature and volume of gas, which have adverse effects on the performance of ESP. Temperature has an adverse effect on the performance of electrostatic precipitators. As an example, magnesite kiln where the kiln outlet temperature is 550–600°C or cement kiln where the temperature is about 350°C may be taken. The use of evaporative cooling in a cooling tower is very common to make the dust more conductive to be treated in an ESP (due to resistivity problems). There is a possibility of wet bottom (slurry) carrying excessive moisture with the dust or the noncontainment of desired temperature that results in operational difficulties. Moreover, due to excessive moisture, dust may harden (like in the case of cement kiln or magnesite kiln etc.) and forms a layer on collecting, and discharges electrodes that are difficult to dislodge, and therefore suppresses the corona and disturbs current distribution. This necessarily reduces the performance of precipitators. This problem necessitates increased maintenance in the form of cleaning the collection and discharge electrode. Also, the sulfur in fuel, in the heat of combustion, forms SO_2 and sulfur trioxide (SO_3), which, in the presence of condensed moisture, forms sulphuric acid (H_2SO_4), thereby corroding the equipment.

   In contrast, fabric filtration is successful even under very high temperature and under different chemical conditions. In the case of treating kiln gases, fire and

Table 4. Comparisons between fabric filter, electrostatic precipitators, and wet scrubbers.

Design parameter	Fabric filter	Electrostatic precipitator	Wet scrubber
Operating principle	Mainly by physical means using filter cloth	Utilize electrical energy directly to assist in the removal of particulate matter	Utilize a liquid to assist in the removal of particulates from the carrier gas stream
Outlet emission	Very low emission is easily achieved with proper fabric material	Very low emission is difficult to achieve within a viable size	Very low emission is difficult unless a very high energy is spent
Meeting stringent emission norm	Removal of smaller particle through changing the fabric filter; therefore it is cost-effective	Through application of stronger electrical field and/or large size/greater number of field; therefore more cost in retrofitting/upgradation	Once designed for a particular efficiency, higher performance cannot be obtained with same size, if required
High CO level	No electrical discharge and hence can operate at higher CO level. However, proper safety measures for operating personnel are required	Explosion due to electrical discharge may occur. ESP is shut down at high CO level resulting into very high dust emission	No effect
Particle re-entrainment	Not applicable during offline cleaning	Unavoidable during rapping	There is no particle re-entrainment
Effect on water pollution	Nil	Nil	A separate water treatment plant is necessary to avoid water pollution
Operation ease	Easy to operate	Complicated to operate	Knowledge of chemistry is required
Pressure drop	High pressure drop	Low pressure drop	Very high pressure drop
Dust resistivity	Fabric filters are not sensitive to dust	Efficiency is lowered for higher dust resistivity. Problem will be aggravated at higher dust loadings or flow rates	Not sensitive
Gas temperature	Not sensitive to temperature as long as fabric material is suitable for that temperature	Very sensitive to gas temperature. Electrical properties of gas deteriorate due to higher temperature	They are not sensitive but at high temperature, steam will come out of stack as white plume

(Continued)

Table 4. Continued.

Design parameter	Fabric filter	Electrostatic precipitator	Wet scrubber
Handling of collected dust	They can handle dry dust easily. However, there is a problem when dusts abrade, corrode, or blind the clothes	Dry dust can be handled easily, but the poisonous gas, ozone, is produced during gas ionization in electrodes	Wet disposal of the collected material is difficult to handle and needs elaborated system to avoid water pollution
Scarcity of water	No effect of scarcity of water on fabric filter	No effect on ESP, unless gas conditioning tower is used	Wet scrubbers cannot be used where water availability is low
Corrosion problems	No corrosion problem occurs or little corrosion	None	In wet scrubbers, high corrosion and allied problems occur
Need of downstream equipment	Not required	Not required	Effluent treatment plant is required
Capital cost	In FF, flange to flange cost is higher than scrubber but lower than ESP	In ESP, flange to flange cost is high, so initial cost is high	Scrubber alone is cheaper but together with downstream equipment, cost is comparable
Maintenance cost	High maintenance cost (bag replacement for clogged/damaged bag)	Maintenance cost is nominal, low operating cost, unless corrosive or adhesive materials are handled	Moderate-to-high maintenance cost owing to corrosion and abrasion
Power consumption	Low power consumption	Low power consumption	Highest power consumer
Operating cost	Low	Low	Highest

heat resistant filter fabrics can be used. In the case of high-temperature filtration, outlet air can be recirculated within the plant for energy conservation.

4. An increase in gas volume has an adverse effect on ESP performance. An increase of 15% gas volume will result in a decrease in efficiency and calls for 15% increase in collecting electrode area to maintain the same efficiency. In contrast, a fabric filter performance is not adversely effected by the change of inlet parameter and this is the reason that a fabric filter is termed as a constant emission machine.
5. A fabric filter system is very effective for the collection of fine particulates and metals. With additional dry sorbents, air toxics, such as Hg, can be collected with substantially reduced injection rates of sorbents over ESPs. Additionally, when following dry flue gas desulfurization (FGD) systems, additional SO_2 capture takes place across the filter cake of the fabric, which enhances the total reduction across the system and lowers consumption rates over a ESP system.
6. The plants with ESP systems emitted more Hg^0 than Hg^{2+}, while the plants with the fabric filters (FF) emitted less Hg^0 than Hg^{2+}. It may be added that the Hg in flue gas mainly existed in the forms of Hg^0 and Hg^{2+}. Virtually all of the Hg^P entering the ESP or the FF was removed. Overall, the FF systems had better Hg^0 and Hg^{2+} removal efficiencies than the ESP systems. In a study, it was found that the average Hg-removal efficiencies of the ESP systems was 11.5%, that of the FF systems was 52.3%, and that of the combined ESP + FGD systems was 13.7%, much lower than the combination of FGD with fabric filter [44]. However, on a per-unit-pressure-drop basis, Hg^0 uptake within an ESP exhibited better performance than the fabric filter, particularly for the low-capacity sorbent and high mass loadings of both the sorbents [55].
7. In an advancement, catalytic fabric filters are found to be capable of removing particulates from flue gases (e.g. from waste incinerators, pressurized fluidized bed coal combustors, diesel engines, boilers, biomass gasifiers, etc.) and simultaneously abating chemical pollutants (e.g. nitrogen oxides, dioxins, VOCs, tar and carbonaceous material, etc.) by catalytic reaction.
8. Trace organic emissions from municipal waste combustors (MWCs) are dependent on the combustion and flue gas-cleaning technologies employed and the conditions at which they are operated. Good combustion practice and dry scrubbing techniques employing fabric filter baghouses can be used to reduce the average polychlorinated dibenzo-*p*-dioxins and dibenzofurans (PCDDs/PCDFs) emissions to less than 20 ng/dscm. Dry scrubbers in combination with ESPs are less effective, resulting in typical PCDDs/PCDFs emissions of less than 75 ng/dscm. Powdered activated carbon can be injected into the flue gas to reduce Hg emissions to less than 0.08 mg/dscm. It can also be used to improve PCDDs/PCDF capture in these systems [56].
9. ESP is believed to give less running cost in terms of power consumption than a fabric filter, but a closer look may give a different perception. On the other hand, a scrubber, while controlling air pollution, creates water pollution. The slurry coming out of the scrubber is difficult to handle and dump. Therefore, cost of downstream system when added to the cost of basic scrubber will not make the scrubber system cheaper. Before taking up the costing of various equipments, it may be worth making a comparison of the control equipments.
10. Collected material is dry for easy subsequent handling and treatment.
11. Corrosion is not an issue.
12. No high-voltage hazards.

13. Many filter configurations available to fit wide range of specifications.
14. Simple operation.

However, fabric filters possess some disadvantages [16,53,57–66], as listed below:

1. Lower temperature limits for standard fabrics.
2. Fabric filters are usually quoted as being large and expensive to build, and in certain cases, unreliable in operation due to their sensitivity to dew point and temperature excursions. There is also a tendency for pressure drop to increase over the life of a set of filter bags. However, operational problems can be minimized/overcome by suitable design of filter material and through selection of appropriate operating parameters.
3. Baghouse design can affect bag movement, which causes the mechanical failure of filter bags [61]. If bags can swing and rub against each other, bag abrasion is likely to occur. Significant bag movement can bend the bag support cages, create sharp sections over which the bags could rub on [16], and possibly even puncture the bags. However, these problems can be minimized/avoided by suitable filter unit design.
4. For certain types of dust, e.g. abrasive dust, very hot particles, light and fluffy dust, mixture of solid and liquid particles, particles causing fire hazards/explosion, particles with strong adhesion forces, etc., proper filter design and special fabric finish are required.
5. High gas temperatures with acid or alkaline gases or particulates can shorten filter life.
6. Moisture or tar in the collected particles can require special cleaning procedures.
7. Medium pressure drop requirements.
8. Replacing fabrics usually requires respiratory protection.
9. Relatively high maintenance requirements.

The decision to use baghouses over other particulate control devices, such as ESPs, depends upon several factors. These include the following:

- Application;
- Acidity of flue gas stream and materials being collected (i.e. high sulfur coals);
- Particulate/ash resistivity (ease in collection in an ESP);
- Efficiency of collection required;
- Operating temperatures;
- Requirement to collect metals and toxins, such as Pb, Cd, Zn, dioxins, HCl, and Hg;
- Collection efficiency for fine particulates (PM_{10}, $PM_{2.5}$).

Variation in the acceptability of fabric filters was also reported. To clean dust-laden fumes from utility boilers, ESPs are generally characterized by higher capital investments and lower operating charges, while the opposite may be said for FF baghouses. Therefore, ESP presents higher total costs when high specific collection areas are required, as happens in the case of low-sulfur, high-resistivity dust. However, significant reductions in both capital investment and operating charges have been evolved in the case of ESP with pulsed energization of precipitators working in severe back corona conditions. This possibility greatly enlarges the field of applications in which ESPs become a lower cost option compared to fabric filters. In a study [67], an economic comparison of pulse-energized ESP with conventional ESP, reverse-air, shaker, and pulse-jet baghouses was made. A mapping of the operating conditions was also put forth to find out economically convenient control technologies. In a separate study, the reverse-flow gas cyclone system showed a better performance than an online pulse-jet bag filter, and substantially better than multicyclone systems under certain circumstances with recirculation. The generally observed unexpected high collection of submicron particles, which occurs with inlet concentration as low as 100 $\mu g/m^3$, is attributed to turbulent dispersion either by promoting fine particle capture by larger ones, much like what occurs in recirculating fluidized beds or by bringing fine particles near the cyclone wall [68]. However, in generalities through sustainable development, the degree of particulate control normally found in bag filter is unheard of in any other type of filter of similar duty.

Declining industrial growth, pressure of cutting costs from every corner, lower price realization, reducing bottom line, restricted capital investment coupled with concern for the environment, forced all concerned, from user to manufacturer, to think of low, cost-effective, and efficient alternative solutions for pollution control. Moreover, pollution control is no longer a low priority area. Industries are eager to see their equipment work best to achieve a clean environment, which also contributes to plants, overall efficiency, and productivity. It is desired to have equipment that is more efficient to achieve even lower emissions than the stipulated standards. The restrictions on one hand and desire on the other, lead technologists and manufacturers to think of innovative ways of upgrading the existing control equipments. Economics of such innovations, however, are important to choose amongst alternatives. Thus, before the existing control equipment is discarded, a judicious analysis is required to justify the cost of abandoning it. It is often possible to upgrade the old equipment to keep the cost low on the one hand and meeting the regulatory norms on the other. The technique employed is generally known as *retrofitting* (*retro* means *backwards*; *fitting* means appropriate). Thus, *retrofitting* means *past conditions appropriately changed*. Figure 5 shows retrofitting the existing ESP casing with pulse-jet fabric filters (pulse-jet low-pressure, high-volume type). Upgrading may be required when

- existing control equipment is not performing to the desired level;
- existing control equipment becomes unsuitable due to change in processes like a dry process to a wet process in a cement manufacturing unit; and
- introduction of more stringent emission regulations.

It is important to note that, in the fabric filter, three basic systems, such as shaker, reverse-jet, and pulse-jet, are used. From the late 1980s, use of industrial bag filters operated in the principle of pulse-jet filtration has got rapid surge, as it proves to be most efficient and versatile. Cleaning occurs more frequently than reverse airflow cleaning, and furthermore, with the use of nonwoven media, much higher levels of air velocity have become achievable than reverse jet filtration. Consequently, it is possible to achieve increased filtration capacity within a smaller area. Furthermore, emission levels are lower in contrast with reverse jet cleaning systems. Cleaning is, in general, accomplished online, but offline cleaning is also employed in very specific cases. Extra bags may not be necessary to compensate for bags during cleaning. Movements of bags are less during cleaning, and can be placed more closely together. As these require less space, they can be installed in limited space at a lower cost than the other system. However, the cost saving may be somewhat counterbalanced by the greater expense and more frequent replacement requirement of bags, the higher power uses that may occur, and the installation of fabric filter framework that pulse-jet cleaning requires [22]. If the upgrading of air pollution control equipment in old plants involves replacing an existing electrostatic precipitator, the high-ratio filter (higher air-to-cloth ratio) based on pulse-jet cleaning can often be installed in the precipitator casing, utilizing the existing support structure, which reduces the investment cost considerably.

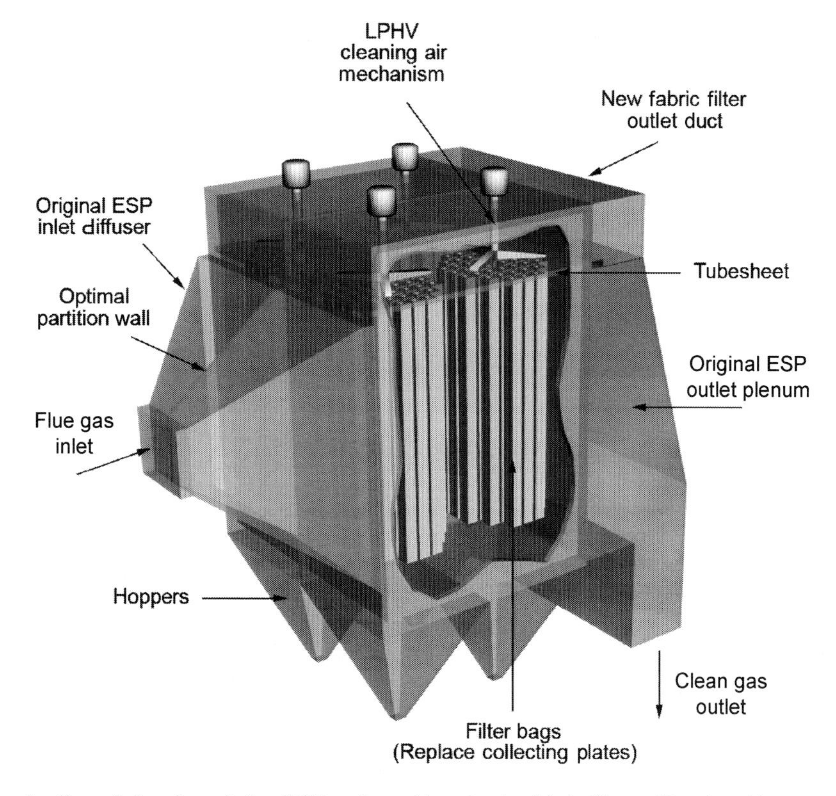

Figure 5. Retrofitting the existing ESP casing with pulse-jet fabric filters. (Reprinted by permission of Hamon Research-Cottrell, Inc., USA.)

High-ratio fabric filter (HRFF) allows flexibility in boiler operation including choices of coals as compared to ESPs [51]. However, ESP and wet scrubbers are also the preferred choice in some specific applications [13,52,69].

Particulate removal is a basic component in advanced power-generation systems such as integrated gasification combined cycle (IGCC), pressurized fluidized bed combustion (PFBC), and other industrial processes to prevent the downstream equipments from fouling and erosion effects [70,71] as well as to meet the environmental regulation limits. Under such situations, rigid ceramic filters, such as filter candles operated under a pulse-jet cleaning system, has emerged as the most promising choice for hot gas cleaning due to its high-filtration efficiency and ability to withstand attack by aggressive gas at high temperature [72–74].

2.2. *About pulse-jet filter*

Pulse-jet fabric filters (PJFFs) are widely used in power generation, incineration, chemical, steel, cement, food, pharmaceutical, metal working, aggregate, and carbon black industries [63]. Figure 6 shows a typical set-up for industrial pulse-jet filter units. With the increasingly stringent emission regulations of fine particulates and air toxins, pulse-jet fabric filters have become an attractive particulate collection option for utilities. Pulse-jet filtration started its debut almost at the same time as the reverse air or shake deflate cleaning arrangements found their use in coal-fired boilers. In a report published in 1992, Belba et al. [75] presented

Figure 6. A typical set-up for industrial pulse-jet filter units (*developed by Hamon Research-Cottrell, USA*).

a worldwide data on pulse-jet filters. It was reported that 300 pulse-jet fabric filters were installed across the world in industrial and utility coal-fired boilers, the first of which was installed for collection of fly ash in the early to mid 1970s. Fabric filters with pulse cleaning have been successfully applied downstream of all kinds of boilers ranging from pulverized coal-fired, stoker-fired, bubbling and circulating fluidized bed combustion boilers. Almost all utility power plants around the world have successfully utilized the fabric filter with pulse-jet cleaning using needle felt fabrics. There has been a definite increasing trend toward the application of pulse-jet filters to larger utility boilers [13].

In a study [37], it was found that, in case of high-ratio fabric filters using the fabric of polyphenylensulfide (PPS) with an intrinsic Teflon (PTFE) coating, particle emission was less than 15 mg/Nm^3 during normal boiler operation for more than 31,000 hours. The particle mass size distribution at the fabric filter outlet showed significantly lower fractions of $PM_{1.0}$ and $PM_{2.5}$ compared to the data from ESPs operating at similar emission levels of boilers fired with coals of different origins and biomass. This suggests that a small diffuse leakage through the fabric filter contributes to the emitted particles. The mass fractions of $PM_{1.0}$, $PM_{2.5}$, and PM_{10} at stack conditions were around 5%, 15%, and 80%, respectively. The fabric filter collection efficiency as a function of particle size was approximately constant at 99.5% in the size range of 0.2–3 μm. The concentrations of the trace elements arsenic (As), cobalt (Co), copper (Cu), lead (Pb), and antimony (Sb) were low in the $PM_{1.0}$ fraction of the emitted particles. High Hg-removal efficiency of more than 80% was obtained in the high-ratio pulse-jet filter in two tests without any injection of an adsorbent [37].

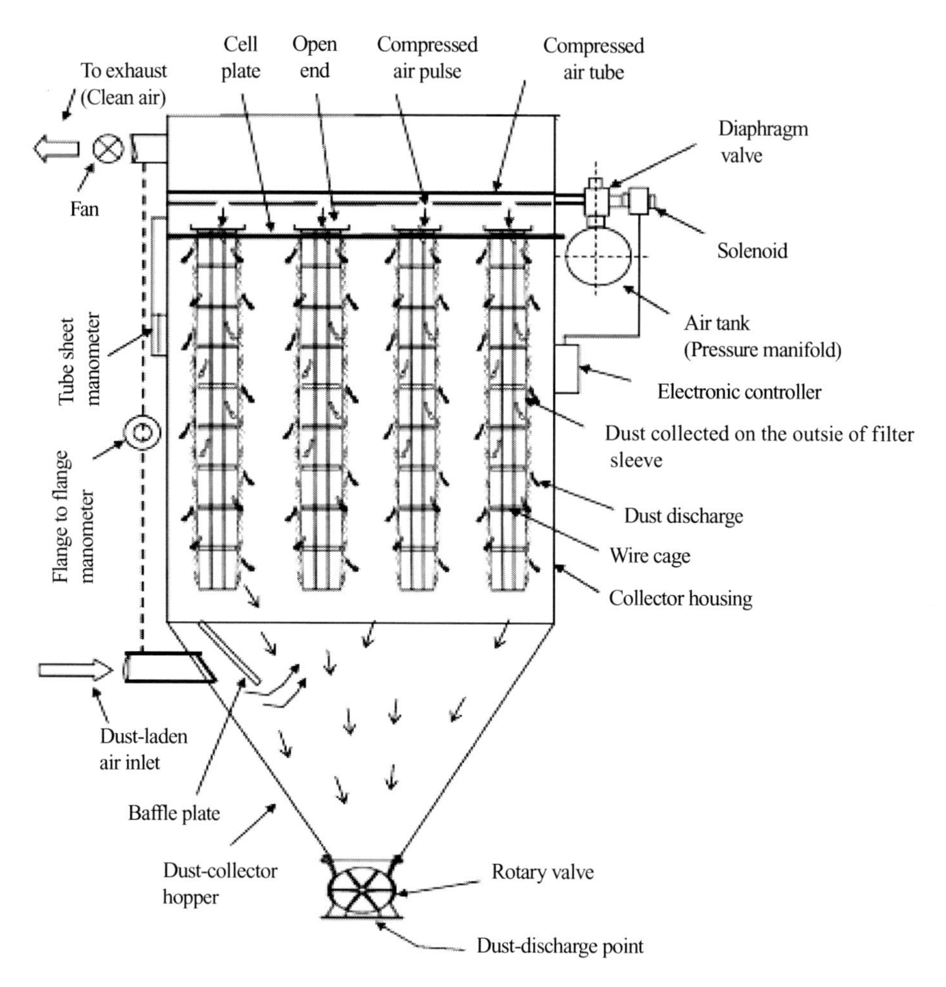

Figure 7. Hardware of bag filter.

The basic hardware of pulse-jet fabric filters is shown in Figure 7; different types of bags are shown in Figure 8. During filtration, gas passes from the outside to the inside of the cylindrical bags, held open by interior metal cages. The particles deposit on the external surface of the filter bags, thus allowing the clean air to pass through. Cleaned gas passes through the top of the bags. With the particle deposition, a positive effect is associated with greater filtration efficiency due to cake filtration, and on the other hand, a negative effect is associated with increased pressure drop. Since industrial filters encounter high-dust density (more than 250 g/Nm^3), pressure drop increases steadily with time. Therefore, filter bags must be periodically regenerated, usually by pulse-jet cleaning. This operation involves injecting high-pressure back-pulse air (3–7 bar) into the filter bags for a very short time (50–150 ms). Back pulse-air injection causes sudden expansion of the filter medium and drives deposited particles and agglomerates away from the bag surface into the dusty gas stream, from which they fall into the dust hopper. Many bags operate in parallel inside the filtration unit and are cleaned in groups (placed in a row) by a momentary pulse of compressed air introduced at their tops. Usually, the bags are cleaned online, that is, at the time of cleaning of a row of bags the process gas flows through other rows. During

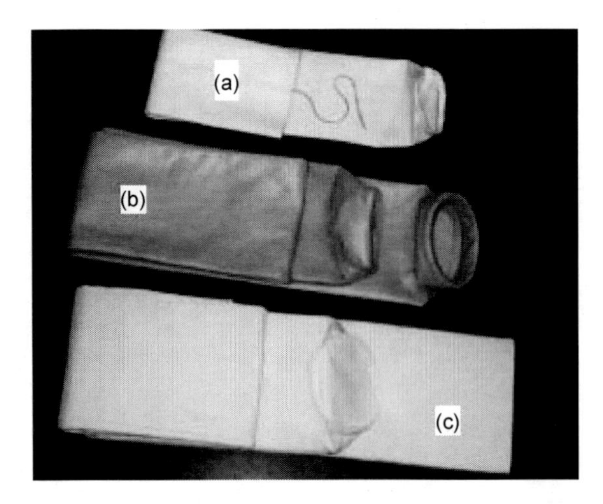

Figure 8. Different types of bags. (a) Plain top, disc bottom with ground wire; (b) Snap band top, disc bottom; (c) Plain top, disc bottom.

online cleaning, the back-pulse is of very short duration and the air filtration is continuously maintained.

During pulse cleaning, the particles retained by the fabric are to be removed either at the upper limit of pressure drop or at a pre-set filtration time to sustain a semicontinuous filter operation. Usually the process plants, and hence the filter bags, operate at steady gas flow. The gas flow is maintained by a downstream fan, which provides the necessary head to overcome the increasing pressure drop across the filter. For the uninterrupted process operation, understanding of the dust layer formed over the fabric and its interaction with fabric structure is necessary. Properties of dust layer/cake depends on many factors such as filtration velocity, dust concentration, etc. [76–80].

The unique features of pulse-jet filter can be summarized as follows:

1. Pulse-jet filtration can meet the stringent particulate emission limits regardless of variation in the operating conditions. It is unaffected by dust variations and provides enhanced gas absorption capacity with fine particle separation. The cleaning device is less expensive than other types of mechanism and requires considerably less space.
2. Pulse cleaning allows operation at high face velocity due to the use of nonwoven fabrics (offers higher porosity as compared to woven fabric used in reverse jet/mechanical shaking system) and less resistance offered by dust layers formed over the fabric. For example, a filtering velocity of 0.02 m/s in felted fabric gives the same pressure drop as that for a filtering velocity of 0.015 m/s in woven glass fiber fabric. It is also observed that, for a given filtering velocity, flange to flange pressure drop was lower for pulse-jet filter in pulverized coal-fired application than for sonic-assisted reverse air or shake deflate filter [81]. For this reason, a smaller number of bags with smaller bag dimensions can be used, which implies a smaller size of filter unit. Further, low filter pressure drop during filtration results in lower fan energy consumption.
3. It has few moving parts, usually a simple solenoid-actuated valve controlled by an electronic timer rather than mechanical shaking motion linkages or large control valves for compartmentalizing and directing flow. However, pulse-jet filters are also

available, which work offline at low pulse pressures. Bags can be more closely placed because of less movement of the bag during cleaning.

4. Duration of cleaning is insignificant compared to the length of filtering time between cleaning intervals. It also offers low operating costs due to energy efficient pulse-jet filter-cleaning processes. However, higher filtration rates combined with higher cleaning intensity may result in the reduction of bag life in comparison to reverse gas designs. Furthermore, since the filter fabric undergoes mechanical stress during cleaning, the fabric should have higher dimensional stability. Dimensional stability of fabric can be enhanced by proper choice of material, operating parameters of filter unit, finishing, and through the introduction of scrim.
5. Usually the process does not allow formation of filter cake; cloth resistance can be maintained at a nearly constant value in contrast to other types of filtration processes. In the case of pulse-jet filtration with cake formation, thickness of filter cake is usually smaller than the cake formed in the case of a reverse jet system.
6. Particulate collection mainly depends on fabric structure. High-cleaning efficiency can be obtained using improved nonwoven fabrics. The system provides lower emission levels over reverse gas designs because felt fabrics are inherently more efficient than woven fabrics.
7. Outside cleaning allows the bag maintenance in a clean and safe environment. In a more specific way, it provides ease to plugging of defective bags and easily accessible service area from above, requiring minimum shutdown and service.
8. All emissions are through the small area, no larger than the filter element cross-section as against the entire fabric surface in case of inside-to-outside filtration. Therefore, the control of leakage through the filter element is easy.
9. Jet-pulsed filters are frequently used for effective separation of filter cake imbibed with gaseous pollutants. An example for the application of the filter as both a solid separator and a fixed-bed reactor is a dry flue gas-cleaning process, where the dry solid sorbent, calcium hydroxide ($Ca(OH)_2$), forms a filter cake that captures a major part of the SO_2 and HCl out of the flue gas.

2.3. Classification and utilities

Pulse-jet filtration can be classified in different ways as given below:

Based on mode of filtration

i. Online filtration
ii. Offline filtration

Based on pulse pressure

i. High-pressure
ii. Medium-pressure
iii. Low-pressure

Based on cake formation

i. Cake filtration
ii. Non-cake filtration

Based on the shape of fabric filter

i. Bag filter

ii. Envelope/oval filter
iii. Cartridge filter

Based on media type

i. Flexible filter (fabric filter)
ii. Rigid filter (such as ceramic filter)

The pulse-jet filter may be cleaned either online, i.e. without isolating the bags from the flue gas, or offline. In the case of online filtration, high or medium pulse pressure is used for bag cleaning. The low ratio design (low air-to-cloth ratio) operated at low pulse pressure and the bag must be cleaned offline. The compartments are isolated from the flue gas periodically and sequentially. The baghouse design modifications required for offline cleaning consists primarily of the added ducting and valving required for compartmentalization. Compartment cleaning can be carried out sequentially, each compartment in turn being taken offline (the outlet valve closed), pulse cleaned, settled, and brought back online. During offline cleaning, the filter bags in the compartment do not take active part in the filtration of the flue gas. Although online mode is much more prevalent, in a few research studies [82–84], offline mode is advised to avoid dust redeposition on filter bags.

The cleaning method used for online pulse-jet filters is much more energetic than the cleaning mechanism used for low air-to-cloth ratio filters. This energetic cleaning allows online pulse-jet filter to operate at much higher air-to-cloth ratio than that of a low ratio filter. The differences between high-, medium-, and low-pressure pulse-jet units are discussed in Section 3.2.1. Pulse-jet filtration is usually non-cake filtration due to a short cleaning interval; however, cake filtration is also common. A brief description of different shapes of fabric filters and rigid filters is given in Section 3.2.

The pulse-jet filtration process is also extensively used for simultaneous control of gaseous emission. In general, for removal of gaseous pollution during coal combustion, the following techniques can be used:

- Injection of an alkaline slurry in ($Ca(OH)_2$, $NaHCO_3$) a spray dryer with the collection of dry particles in a fabric filter;
- Dry injection of alkaline material into the flue gas stream with collection of dry particles in a fabric filter;
- Addition of alkaline material to the fuel prior to or during combustion.

Of the above-mentioned systems, spray drier filter systems for the simultaneous removal of SO_2 and fly ash from industrial boilers become very common. Everaert et al. [85] discuss the use of activated carbon to adsorb dioxins and furans, and then to capture the carbon by fabric filtration. The carbon may be injected as particulates into the flue gas stream. Wet FGD and in-duct sorbent injection are typical methods used widely for gaseous Hg removal emitted from industrial facilities. However, wet FGD cannot remove Hg^0 that occupies 50% or more in the Hg species emitted depending on operational conditions. Sorbent injection method has already been applied to incinerators or power plants to remove gaseous Hg in developed countries, because it can remove Hg^0 as well as oxidized Hg. However, since the removal characteristics of gaseous Hg within flue gas by sorbent injection are largely affected by the composition of flue gas and the operational conditions, it is difficult to predict the characteristics of its removal [86].

One of the major technological hurdles with the commercialization of advanced gasification-based power and hydrogen-production systems is an unreliable gas-cleaning process. The raw syngas from the coal gasifier contains a number of solid and fluid contaminants, which have to be removed at the highest possible temperature to achieve optimum cycle efficiency and protect downstream process equipment and catalysts. Existing technology for gas cleaning at high temperature involves candle filters for removing solid contaminants and sorbents for removing fluid contaminants.

Among many design considerations of pulse-jet filters, fabric selection for dust collectors is a vital factor in the air pollution control equipment decision that must be made on a location-by-location basis. Filter bag elements represent a major investment as well as a highly technical decision that warrants extensive evaluation. Apart from the flexible fabric filter, rigid ceramic filter is also used at very high-temperature applications. In order to meet the regulations that have been consistently changing for the last decade, many plants have substantially changed their filtration processes. These process changes have created a change in the type of particulate being collected in the baghouse. This is necessary due to the constantly changing regulations that appear to be heading toward $PM_{2.5}$ collections.

3. Mode of filtration

In the industry, filtration process is executed by collecting the dust-laden air from various operations and passing it through the filtration equipments. The clean air exits from the filtration device. The main function of filter medium is not only to remove particles from the system but also to allow the air to pass through. The main requirements of filtration setup are dust air collection, filtration, cleaning, and collecting removed dust. To execute these functions, various methods are followed, which will be discussed further in detail. In the case of baghouse filtration, the design of filter medium is mostly in the form of tubular, which facilitates the feasible aerodynamic mechanism to promote the effective filtration. Baghouse equipment is classified according to its design, method of cleaning, construction, and size. There are two filtration designs in accordance with the direction of fluid flow [13]:

- Inside-to-outside filtration mode;
- Outside-to-inside filtration mode.

Inside-to-outside filtration mode is applicable in the case of mechanical shaking and reverse airflow filtration systems, whereas outside-to-inside filtration is prevalent in the case of filtration with pulse cleaning. In pulse-jet filtration, the filter element can vary from bag filter to rigid ceramic filter. Construction of baghouse differs depending on the mode of filtration adapted in different industries.

3.1. Inside-to-outside filtration

In this case, dust-laden air enters into the filter element through the inner side, and clean air exits after depositing the dust cake in the inner side of the bag filter element (Figure 9). This mode is prevalent in the case of the reverse-/shaker-type filtration method. The top portion of the bag is kept closed and fixed, and the bottom is designed as an open portion for dust air to enter. However, the above system with a mechanical type of cleaning has some inherent limitations, i.e. the life of the bags used in the mechanical shaker is less because shaking of the bags causes fiber–fiber abrasion and flex fatigue. In the case of cleaning by reverse airflow, ambient air is made to pass through the fabric in the opposite direction to the gas flow. During cleaning, bag tends to deflate or collapse, which restricts the flow to dislodge dust from free falling, and hence anticollapse rings are provided [13,81].

Like mechanical shaking, the system runs in offline cleaning principle and therefore compartmentalization is required. Reverse air cleaning is better than shaking and compatible

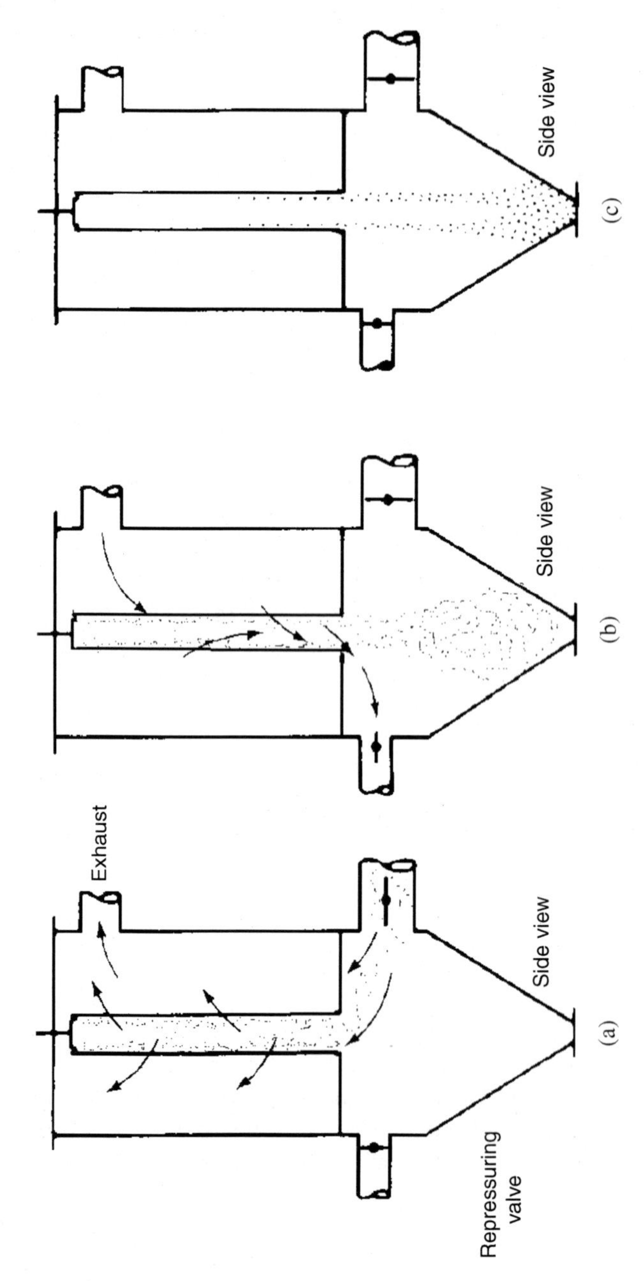

Figure 9. Inside-to-outside filtration. (a) Filtration; (b) Reverse jet cleaning and dust dislodgement; (c) Dust settlement.

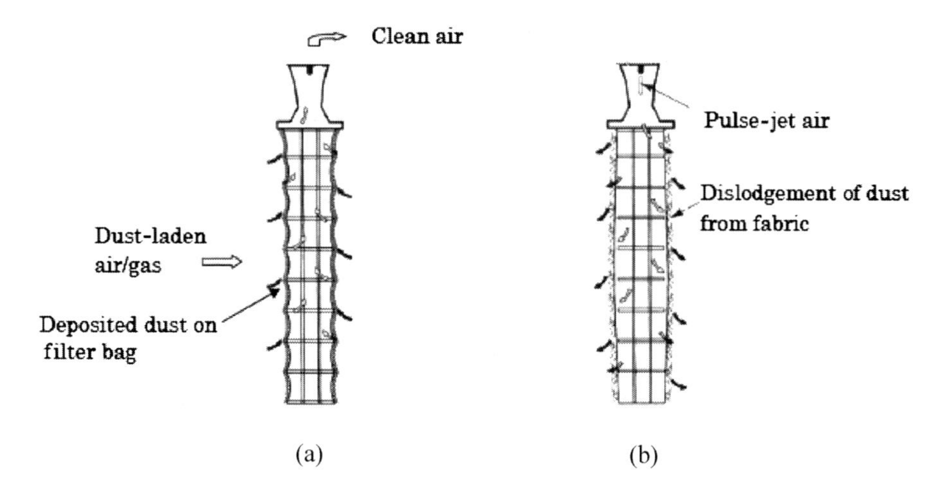

Figure 10. Outside-to-inside filtration. (a) Normal filtering mode; (b) Pulse-jet cleaning operation.

with fiberglass filter bags. Reverse air cleaning is ineffective for needle felts because of its low intensity of cleaning, which is not efficient in the removal of trapped particles from the structure. In the inside-to-outside filtration process, to maintain an acceptable static pressure, the secondary dust cake must be removed on a regular, sometimes continuous basis. However, this excessive cleaning can upset the primary dust cake and emissions are high until the primary dust cake is again built.

3.2. Outside-to-inside filtration

The schematic diagram of the above process is shown in Figure 10. Dust-laden air enters the filter element from the outer side and clean air exits through the top of the filter bag. Thus, dust deposits on the outer side of the bag filter element. This mode is adapted in the pulse-jet cleaning-type filtration method (Figure 7). A pulse-jet unit can be square, rectangular, or cylindrical, small/very large, etc. The top portion of the bag is kept open and fixed, and the bottom portion is closed. The bags are cleaned by a pulse of compressed air, which is fired axially from the top, open end of the bag. Filter bags are cleaned row-wise; once one row of bags is cleaned, the next valve is actuated to clean the next row. A new cycle begins as soon as the pressure drop reaches a certain pre-defined upper limit (Δ P_{max}) or a pre-defined filtration time is elapsed [13,87]. Usually the time interval for cleaning consecutive bags can be set conveniently and the cyclic process continues as per the process plan [13]. It may be added that in a patented technology, onset of pulsing is based on emission [88]. The pressurized air pulses are supplied to the row of bags by a tube provided with small orifices or nozzles, which direct the jets of compressed air at a high velocity into the top of the bags. The distribution of the compressed air in short pulses is performed usually by a solenoid-operated diaphragm valve, connected to a tank containing compressed air [81]. At a pre-determined time period, this valve actuates the pulsing mechanism in order to clean the filter element to keep the system in a steady state of operating pressure drop. This was archived by energizing the diaphragm with respect to the cycle time and deenergizing with respect to the pulsing time. The duration of the pulse is in the order of milliseconds; hence, there is no need to shut the flow of dust-laden air for cleaning purposes. Hence, this system does not require compartmentalization [13].

As the airburst travels down the bag, the normal flow of the flue gas through the bag is stopped, and the pressure and shockwave is transmitted down the length of the bag. Pulse cleaning is carried out with two mechanisms [89], shock pulse wave and rapid reverse flow through the medium. Pulsing of compressed air is opposite to the direction of flow of dust-laden air; hence, the dislodged dust eventually settles down in the hopper located below the bags.

The filter cleaning is a complicated phenomenon, which is influenced by many factors such as the force acting on the filter, the cleaning air volume, the peak pressure, and the pattern of cleaning pressure buildup. During the filter cleaning, the dust cake detaches from the filter surface the moment the force reaches the maximum and a fraction of removed dust reattaches to the filter surface at the end of filter cleaning because of the reversed gas flow towards the filter surface. The filter-cleaning efficiency is therefore affected by the combination of these detachment and reattachment mechanisms. The detachment of particles is enhanced with the peak pressure and the reattachment is reduced with increasing the time of pulse-jet.

Usually outside-to-inside filtration operates in the non-dust cake mode of filtration; however, dust cake mode of filtration is also prevalent. The later mode is preferred by those few intrepid utility/large industrial baghouse operators who have opted for a pulse-jet solution and its high-face velocity as the cost-effective way to control boiler emission. Although the basic form of a filter element is a bag filter, other forms such as cartridge and envelope filters, and ceramic candle filters are also prevalent (Figure 11).

3.2.1. Bag filters

A bag filter is the most common type of filter element made out of nonwoven felt fabrics. The bags are usually sewn with three-needle chain stitch or welded vertical seam. In pulse-jet cleaning, the length of the bag plays a role in the effectiveness of the pulse pressure. To obtain the effective pulse pressure, the bag size is designed in a smaller length of six meters. Filter sleeves in pulse-jet collectors are located in an opening in a cell plate and

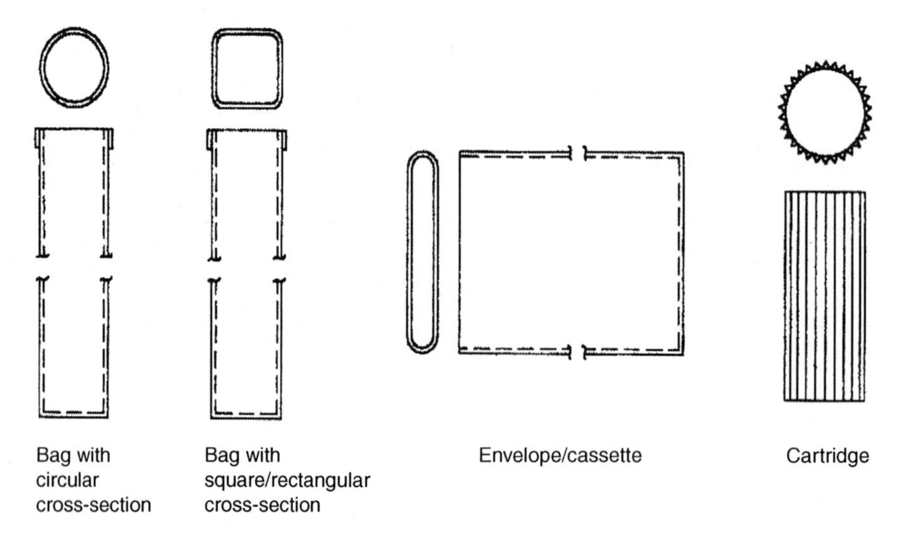

Figure 11. Various forms of filter fabrics.

they may be mounted in either a vertical or horizontal manner. Since dust is collected on the outside of these sleeves, the fixture at the location point is critical if by-pass of the filter and subsequent emission of dust into the atmosphere is to be avoided. Some possible gasketing arrangements are single felt gasket, double felt gasket, spring band profile, and garter springs [90].

Due to higher filtration velocity, the deposition of dust cake takes place at a much higher rate and the further increase results in an increasing pressure drop in a shorter period of time [13]. Depending on the tank pressure employed during pulse cleaning, this system can be classified as follows:

- High-pressure low-volume pulse (HPLV) – Primary pulse of 600 Kpa.
- Medium-pressure medium-volume pulse (MPMV) – Primary pulse of 200–250 Kpa.
- Low-pressure high-volume pulse (LPHV) – Primary pulse of 100 Kpa.

The design of equipment varies with respect to its pulse pressure. Table 5 provides the details of pulsing information for different types of pulse cleaning. The schematic representation of three types of pulse-jet filtration systems is shown in Figure 12.

In the case of high-pressure pulse, air is assisted by venturi, which acts as a jet pump and helps in inducing additional air by six to seven times the jet flow. On the other hand, medium-pressure pulse cleaning is based on the elimination of secondary air and, hence, venturi. The air is controlled by a suitable blowpipe with a 20-mm nozzle for a bag diameter of 120–130 mm. Low-pressure pulse operates at a primary pressure of 100 kPa. In this arrangement, the most common design is to have filter elements of oval cross-sections arranged in concentric circles and using rotating pulse manifold arms. Oval-shaped filter bags are claimed to provide better snap during cleaning than round bags, thus allowing better cleaning. Furthermore, in the above case, the amount of bag material in a given area can be increased. The pressurized gas from an external tank is ducted through a revolving arm, which stretches radially across a pie-shaped segment of a tube sheet separating a clean gas plenum from a dirty gas plenum in a bag collection gas filtration system. The ducted gas is intermittently discharged from the rotating arm through multiple orifices above the bag top. A nozzle cross-section is also rectangular, which is 10% of the filter element cross-section [13,91,92]. A positive displacement blower is activated to pressurize an air reservoir. The design incorporates one large valve and a single rotating blowpipe with large openings. A principle design is shown in Figure 13.

Some investigators [93,94] have compared the cleaning performance of the three above-mentioned commonly used pulse-cleaning designs, i.e. low-, medium- and high-pressure baghouses. It was found that the low-pressure system design without a venturi injector requires lower cleaning energy for an equivalent cleaning efficiency. Humphries and Madden [95] found that particles would redeposit on the bag after cleaning when online cleaning was utilized. Lu and Tsai [96] evaluated two different cleaning types – high initial tank pressure and one bag cleaning, and low initial tank pressure and two consecutive bag cleanings – with regard to pilot-scale, pulse-jet baghouse performance in an oil-fired boiler. Between the two above-mentioned cleaning types, the second one is found to be more effective to clean the bag and the energy consumption due to significant reduction in the use of compressed air as compared to the first type.

In many cases, traditional high-pressure pulse using venturi has been replaced by medium-pressure pulse without venturi (Figure 14). In the latter case, pulse timing is enhanced to achieve effective cleaning, which is based on a medium-volume–medium-pressure concept. The cleaning energies of the low- and intermediate-pressure pulses

Table 5. Differences in filter unit and process parameter based on category of pulse pressure.

Category of pulse pressure	Primary pulse pressure (tank pressure) (kPa)	Pulse pressure (kPa)	Valve size (mm)	Nozzle size (mm)	Bag diameter (mm)	Bag length (m)	Ratio of secondary air to pulse flow	Number of bags per valve	Position of gas inlet
High-pressure pulse	More than 600	3–4	25–65	6–10	115–150	5–7	6–7	Less than 20	Bottom/top/side
Medium-pressure pulse	200–250	4–5	80–100	18–20	125–150	5–8	1–2	10–20	Top/side
Low-pressure pulse	55–100	2–3	150–300	—	75 × 150 mm (race track/oval cross-section)	6–8	—	1000	Bottom straight/ tangential

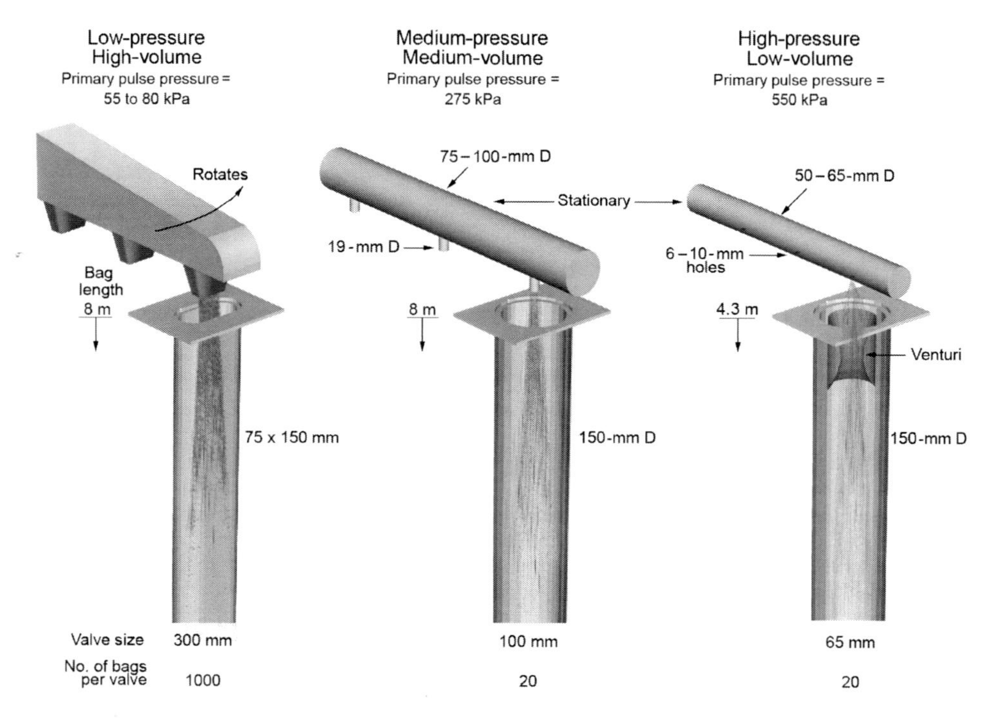

Figure 12. Schematic representation of three types of pulse-jet filtration system. (Reprinted by permission of Hamon Research-Cottrell, Inc., USA.)

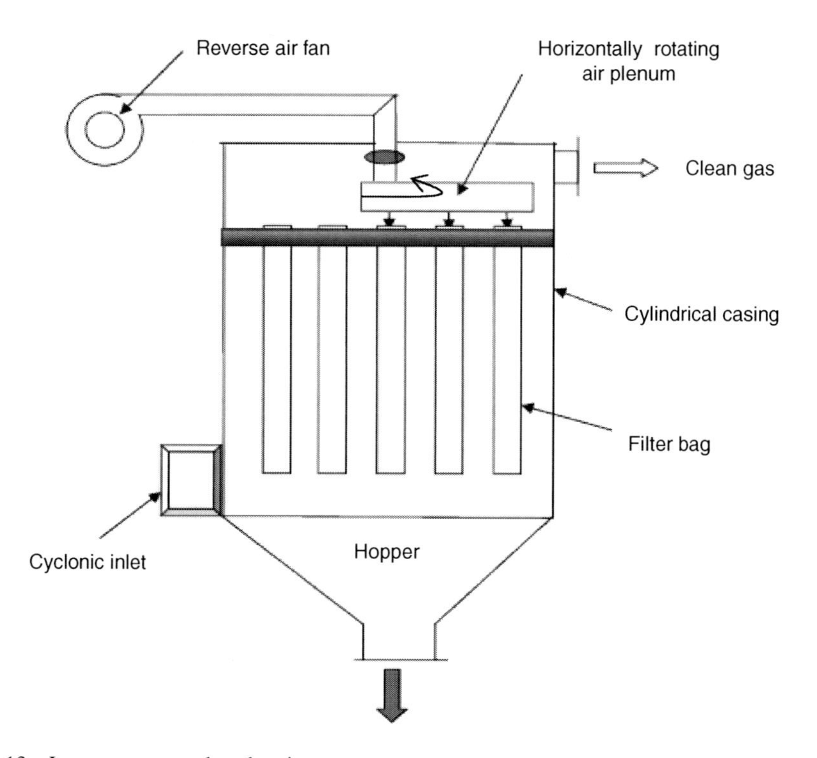

Figure 13. Low-pressure pulse cleaning.

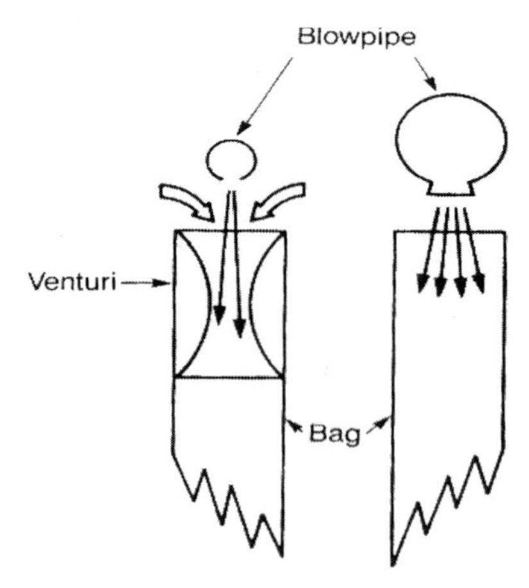

Figure 14. Pulse cleaning with and without venturi.

carry further down the bag than the energy of the high-pressure pulses does. Therefore, high-pressure systems typically have shorter bags than the low and intermediate cleaning technologies. Low- and intermediate-pressure designs are currently under operation with lengths of 7- and 8-m, compared to the typical maximum high-pressure length of 5 m. With the long bags cleaned online, the number of hoppers, as well as the filter support structure, is reduced to cut down ash-handling and civil works' costs. Furthermore, a small footprint makes ideal retrofits in old plants, where there is limited available space between the boiler house and the stack [97]. However, the bag length and cleaning differences were under further scrutiny as a result of fabric filter installations utilizing high-pressure pulsing and 7-m lengths becoming available [98].

As compared to the above system, the same size low-pressure pulse-jet filter will require a positive displacement blower (instead of higher horsepower compressor) and no drier. The air on the medium-pressure unit is not compressed and heated enough to cause condensation. In addition to the considerable level of energy savings, the power required to heat high-pressure compressed air cleaning headers in cold climates to prevent freezing is not necessary. In addition, low-pressure pulse-jet units offer simple design and may only have one or two large valves, which may greatly reduce maintenance costs and still provide the advantage of less energy consumption [99]. A low-pressure high-volume pulse-jet system is suited to large air volume applications collecting powder. A large gas volume (e.g. over 40,000,000 m^3/hr) from various plants can be conveniently dedusted by means of low-pressure pulse-jet fabric filters. A low-pressure pulse-jet has numerous advantages, such as being capable of handling high raw gas dust load, low investment charges, high-collection efficiency, short delivery and erection period, and low operating costs. The system also provides a constant process airflow and fewer moving parts, which in turn results in reduced maintenance and disruptions to system airflow. The LPHV fabric filters are applicable to utility coal-fired boilers (and other large industrial applications) as primary fly ash collection, or in polishing filter services, or downstream of spray dryers. LPHV pulse-jet is a competing technology to the medium-pressure pulse-jets. Hamon Research-Cottrell, USA, predominantly offers LPHV pulse-jet, whereas Siemens, Alstom, and Babcock & Wilcox offer medium-pressure pulse-jets for similar applications. These three firms compete for fabric filter projects and include their fabric filter technology in larger bundled projects such as upstream of their wet FGD scrubbers. Table 6 shows the uses of different cleaning systems' trends in different applications.

Table 6. Cleaning systems' trends in different applications.

Applications	Typical filter Velocity m/min (ft/min)	Typical particulate Loading (g/m^3)	Cleaning systems' trend	Type of cleaning*
Utility applications				
Primary fly ash collection	1.1–1.4 (3.5 to 4.5)	2–25	LPHV/MPMV	Online
Polishing fly ash collection (downstream of ESP)	1.4–2.0 (4.5 to 6.5)	0.6–1.7	LPHV/MPMV	Online
COHPAC (downstream of ESP)	>2.4 (>8)	0.6–1.7	LPHV/MPMV	Online
SDA/FF (downstream of spray dryer)	1.1–1.4 (3.5 to 4.5)	10–35	LPHV/MPMV	Online
CDS/FF (downstream of circulating dry scrubber)	0.8–0.9 (2.5 to 3)	110–460	LPHV/MPMV	Online
Industrial applications				
Steel	1.2–1.8 (4 to 6)	2–12	MPMV	Online
Cement	0.8–1.5 (2.5 to 5)	20–460	MPMV	Online
Smaller volume applications (many industries)	—	—	HPLV	Online

* Offline cleaning is atypical as the base-cleaning mode but can be incorporated in cleaning programs as an option.

3.2.2. Rigidized/cartridge media

To produce a filter bag that is able to operate without a support cage requires a filter medium of much higher flex strength than the conventional fabric. Rigidizing technology is based on thermal compression of conventional needle felts with or without thermoplastic adhesion of fibers, resin-treated nonwovens, coated fabrics, spun-bonded, and sintered polymer constructions. Sintered polyethylene is a typical construction giving limited thermal and chemical durability [100,101]. The term 'rigidized media' was devised by Smith [102] to identify a category of dust filters that evolved from the conventional pocket fabric filter. The rigidized filter media is finely pleated (Figure 15), thereby increasing the filtration area per unit volume by up to 300% depending on existing filter bag sizes. The rigidized media is also sometimes referred to as cartridge media or extended media filter. A typical cartridge collector is approximately four times smaller than baghouse designs for similar gas streams. It has a tight pore structure and rigid physical properties that allow it to hold a pleat without any need for supporting back material. Cartridge filters can therefore be considered as being similar to cageless bag designs, except that some cartridge designs include supporting elements, end caps, perforated tube inserts, and/or pleat support rods of different materials including polymers and metals [100,103]. It can treat high-dust loadings, and concentration of outlet emission is greatly reduced, especially for sintered media. However, the cartridge collector cannot be operated at higher airflow rates. At high-temperature applications, it seems likely that the use will be limited to small-scale incinerator applications [21,100].

Cartridge filters are mostly based on spun-bonded nonwoven media, but filter media are laminated with a membrane with very fine openings for enhanced removal of submicron particulate at the level of 99.999% [104–106]. Such filter media tend to be more expensive. Nonwoven media is generally made from synthetic materials such as Nomex, polyester, or

Figure 15. Cartridge filter.

polypropylene. Some filter media, such as cellulose paper filters, are useful only at relatively low temperatures of 95 to 150°C. For high-temperature flue gas streams, more thermally stable filter media, such as nonwoven made out of Nomex fiber, can be used [103,107]. Among a wide variety of cartridge designs and dimensions, typical designs include flat panels, V-shaped packs or cylindrical packs. Commercially available cylindrical packs are approximately 15 to 35 cm in diameter and 40 to 122 cm in length. The cartridge is closed at one end with a metal cap. The media is sealed with the cap using polyurethane plastic, epoxy, or other commercially available sealant. A neoprene or silicone gasket seals the frame to the clean air plenum of the collector [103]. The cartridge can be mounted vertically as a nearly direct replacement for standard bags and cages in existing baghouses, or mounted horizontally in original designs. Operating conditions are important determinants of choice for filter media and sealants used in the cartridges.

Pleat spacing is important for the following two reasons: closer spacing increases filter area for a specific cartridge volume, whereas closer spacing increases the likelihood of dust permanently bridging the bottoms of the pleats and reducing available filtering area [103]. Pleat arrangement and available volume of cleaning air determine the cleanability of the media for a specific dust [107]. Corrugated aluminum separators are often employed to prevent the pleated media from collapsing. The pleat depth can vary anywhere between 2.5 cm and 40 cm. Pleat spacing generally ranges between 12 to 16 pleats per inch, with certain conditions requiring even lesser pleats, i.e. four to eight pleats per inch [21,103,104]. There is a claim of deep pleat cartridges with unique combination of staggered dimpling on the filtering side, which, unlike other dimpling patterns, provides unrestricted airflow and pulse cleaning at the extreme ends of the pleats. Dimples are formed on thermoplastic nonwoven fabrics with a heated die and staged cooling to set the shape permanently [108].

Pleated filter elements provide a simple retrofit for upgrading existing dust collection systems. Inlet abrasion is a common mode of filter bag failure in the rock dust industries. Installation of shorter filter elements keeps the filters away from the 'abrasion zone', resulting in improved performance and longer filter life. Pulse-jet-cleaned cartridge filters can only operate as external cake collection devices. When the particulate-laden gas flows into the cartridges, diffusers are often used to prevent oversized particles from damaging the filter media; and smaller size particles deposit outside the filter media. During pulse-jet cleaning, a short burst, 0.03 to 0.1 seconds in duration, of high-pressure, 415 to 830 kPa, air is injected into the cartridges. The pulse is blown through a venturi nozzle at the top of the cartridges and establishes a shockwave that continues onto the bottom of the cartridges. The wave flexes the filter media, dislodging the dust cake.

The number of cartridges utilized in a particular collector is determined by the choice of air-to-cloth ratio, or the ratio of volumetric airflow to filter media area. The selection of air-to-cloth ratio depends on the particulate loading, particulate characteristics, and the cleaning method used. A high particulate loading will require the use of a larger number of cartridges in order to avoid forming a heavy dust cake, which in turn results in an excessive pressure drop. Determinants of cartridge collector performance include the filter media chosen, the cleaning frequency and methods, and the particulate characteristics [109].

A typical cartridge contains an inner supporting core surrounded by the pleated filter medium and outer supporting mesh. The cartridge is held tightly in place against the mounting plate surrounding the hole that connects it to the clean air plenum. Horizontal cartridges are typically mounted in tandem with a gasket seal between them. If not mounted properly or if the gasket material is not of high quality, leakage will occur after repeated cleaning pulses [107]. In the case of cartridge filter, quality control must be higher than for conventional bags. The bags are expensive, may fail in fatigue at the pleat, and need high-pulse air to clean the dust trapped in the pleat. Stability and appropriate geometry of pleats are very important for the effective functioning of cartridge filter.

3.2.3. Envelope filter

By using rectangular shape envelopes or flat bags, substantial media area can be accommodated into a relatively small space (Figure 16). Multi-bag envelope design filters have also been used for smaller filters and are characterized by a number of flat bags sewn side to side to decrease spacing even further. Major concerns relating to envelope bags are cleanability as regard to agglomerating dusts, which do not release well, and bridging of dust between tightly spaced bags. Traditionally, envelope bags are used in smaller collectors or confined to specific areas such as packing houses. Sly Pactecon dust collector [110] offers an envelope filter bag for controlling air pollution under continuous operation with offline cleaning. The system claims to provide twice the collection area of tubular bags, longer bag life, and reduced energy consumption.

3.2.4. Ceramic filter

High-temperature filtration from industrial processes offers various advantages in terms of increasing process efficiency, heat recovery, and protection of plant installations. The attempts to efficiently use thermal energy from coal-derived hot gases in advanced coal-fired power plants, using PFBC or integrated gasification combined cycle (IGCC) processes, have driven the development of collection devices for gas cleaning at high temperatures. In fact, the commercialization of these technologies was critically hindered by the problems with the dust removal system operating at high temperature and pressure [111,112]. In modern power plants that utilize IGCC and PFBC, dust and char particles are removed, often under high-temperature and pressure conditions, by closed-end ceramic composite cylinders arranged in a rosette enclosed in a plenum [113]. A few key applications of ceramic elements also include waste incineration, waste pyrolysis, secondary ferrous metals smelting, precious metal recovery, thermal soil remediation, fluidized bed metal cleaning,

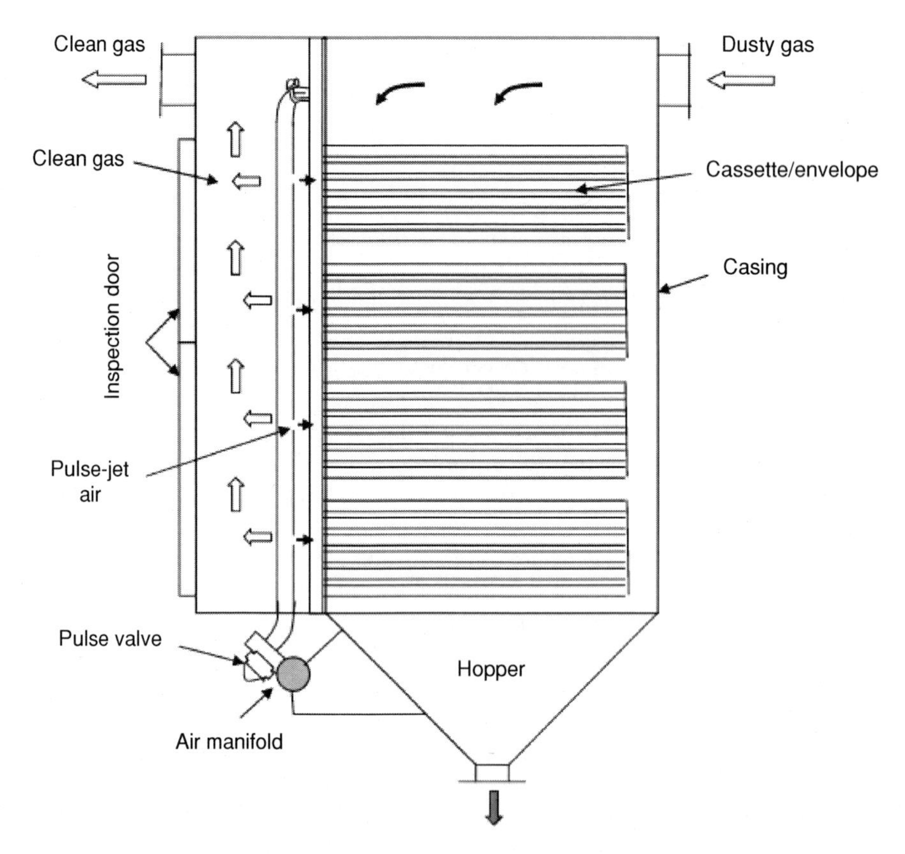

Figure 16. Filtration unit with cassette/envelope fabric filter.

boiler plants, chemicals manufacturing, and glass melting. These can remove and/or recycle catalysts from any catalytic reaction system and strip catalyst from fluid catalytic cracking flue gas. These can also remove contaminants and/or recover valuable products in applications involving halogenated hydrocarbons, petrochemical processing, catalyst activations, fluid catalytic cracking, incarnation of hazardous materials, and combined gasified cycle [114,115]. Filtration of hot gases must be thoroughly performed to protect the downstream heat exchanger and gas turbine components from particle fouling and erosion effects and to clean the gas to meet particulate emission requirements.

A ceramic filter could be a flexible fabric filter or a rigid candle filter. Fabric filters consist of mechanically bonded ceramic fibers, are usually self-supporting, and do not need any supporting cages. Ceramic candles are rigid, closed-end composite cylinders. These filters are very effective for hot gas cleaning owing to its high-filtration efficiency, resistance to attack by aggressive gases, and heat shock resistance. Their potential may be further extended by using them in combination with injection of a dry sorbent to remove acid gases and other chemical contaminants. Commonly, ceramic filter media is an increasingly popular option for particulate removal from hot gas streams at a temperature of about 300°C. A crucial factor in their successful utilization is the ability of the cleaning pulse to remove the deposited cake effectively from the filter surface. In the filtration process, dust cake formed on the filter surface needs to be removed periodically by the pulse-jet cleaning system to recover the permeability of the filter for its successive long-term operation. The pulse overpressure, the pressure difference developed between the inside and outside of the filter during the pulse cleaning, is the main driving force to overcome the binding force between cakes and filter medium and remove the dust accumulation on the surface of the candle filter.

The advantages of ceramic filters can be summarized as follows [100,115]:

- Filtration efficiency can be in the order of 99.99% with a maximum pressure drop of 100 mbar.
- No support cage is required, as the filters are rigid. They can have a significant impact on the capital and running cost of the filter.
- It possesses high safe working temperature, typically 900°C, which is far in excess of the temperatures normally encountered in incinerator/waste heat boiler flue gases. It may also offer considerable thermal protection to the filtration elements in the event of, for example, filter hopper fires, thus adding to the security and durability of the filter.
- Ceramic elements are resistant to acid and alkali species. The chemical stability of ceramic materials is better than polymeric fibers; therefore, these possess an enhanced filter life when compared with polymeric fabrics.

The candle filters are hollow cylinders (closed on one end) typically with a diameter of about 6 cm and a length of 1 to 1.5 m. The thickness of a candle ceramic wall varies between 0.8 cm and 1.5 cm. An industrial filter vessel normally contains a large number (several hundreds) of candle filters. These filters may be made of silicon carbide with silicate binder, mullite with mullite binder, or silicon carbide with mullite binder [116]. Ceramic elements, such as cerafil, are manufactured from mineral fibers, which are bonded together through a combination of organic and inorganic binders. Elements are monolithic and formed into a shape, which incorporates an integral mounting flange resulting in a rigid, self-supporting structure. The manufacturing method, akin to injection moulding, confers uniform physical properties and a tight external surface structure reducing penetration of fine particles inside the structure [114]. In a normal case, a membrane-coated silicon carbide filter candle is used for hot gas cleaning. The filtration membrane is the rigid porous structure, which reduces penetration of dust particles, and hence reduces residual pressure drop across the filter candle. In the high-temperature filtering system, there are many problems which can threaten the stability and reliability of filtration, such as thermal shock caused by using cold pulse-cleaning gas [117], the breaking of the filter element due to mechanical instability [118], bridging of dust cakes between adjacent filters [119], and weakening of the filter-binding force caused by local temperature rise with combustion of the ash accommodated in the filter element [120]. The lifespan of a ceramic filter within a plenum is a function of the manufactured quality of the filter, the thermal stress and pressure cycles, as well as the periodic cleaning of dust deposits using back-pulse flushing. Generally, the utilities of ceramic filters are three years of service at temperatures between 760°C and 927°C. The stability of the filters is critical to many treatment processes, such as high-temperature desulphurization of coal gases [116]. Further, the presence of liquids and formation of molten particulates can hinder the filtration process. This can occur when plant operation is at a temperature at which liquid water can condense onto or into the ceramic elements. Similarly, care should be taken to ensure that the particulate being filtered is not molten when it reaches the filter elements or can become molten or sinter due to excessive temperature excursions [114].

4. Basic construction of a filter unit

4.1. General

In this section, main emphasis is given on the basic baghouse design. The success of filtration equipment depends on the design and layout of proper inlet and outlet, effective extraction and capture of dust, and finally disposal of collected dust. The dust removal device is a fabric, which acts as a barrier to remove particulate from a gas stream. The design system must entail highest filtration efficiency even for smallest particles at lowest possible pressure drop across the system. There are two alternative locations for monitoring the static pressure drop of a fabric filter, i.e. flange to flange and across the tube sheet (Figure 7). The pressure drop across the tube sheet indicates the resistance to gas flow caused by the filter media and the dust cake on the filter media [121].

Higher-than-expected static pressure drop increases the overall system resistance to gas flow. Decreased gas flow from the process area will result if the centrifugal fan and damper system cannot compensate for this increased resistance. Fugitive emissions can occur when the gas flow rate at the hood is too low. High-static pressure drop also increases the electrical energy needed for the centrifugal fan, increasing the operating costs of the system.

Baghouse filtration systems can be broken down as follows:

1. *Dust-laden air-feeding zone*: It consists of capture point and duct work. At capture point, dust-laden air is generated and entrained into an air stream by hoods or enclosures. Ductwork are hollow tube enclosures used to transport the dust-laden air stream from the capture point to the filter unit, and are designed for constant velocity and minimum flow turbulence.
2. *Filter unit*: It consists of filter bag, bag-mounting arrangement and its hardware, and cleaning arrangement (Figure 7). The compressed air required for cleaning should contain minimum moisture and oil so that clogging of fabric is avoided. The limit of 10 gm/m^3 of moisture and 0.01 gm/m^3 of oil could be set for a trouble-free operation of the filter.
3. *Dust disposal system*: Dust collected on the filter element and discharged by the cleaning system falls by gravity into a hopper or bin and is removed periodically or continuously, depending upon system requirements. This can be achieved by mechanical, pneumatic, or manual methods.
4. *Fan (air mover) and discharge stack*: This moves the dust-laden air from the capture point through the ductwork and filter unit by adding sufficient energy to the airflow to overcome resistance to the flow. A fan consists of a bladed rotor (impeller) and a housing to collect and direct the air discharged by the impeller. Discharge stack is the exhaust outlet for the cleaned air system.
5. *Auxiliary attachment*: There could be a number of additional setups for assisting/enhancing filtration process, providing safety for explosive particles, and removal of gaseous emissions. In addition, measurement and control system is necessary to control the process variables as well as to run and monitor the industrial filtration process. Parameters such as temperature, pressure, dust concentration in emission, moisture content, particle size distribution, physical and chemical characteristics of gas, and dust are often to be measured. Further apparatus for bag leakage detection, bag tensioning apparatus, and detection of overload of a fabric filter may also become necessary.
6. *Safety features for explosive and/or inflammable dust*: Designing of the filter equipment is largely influenced by the safety features to be incorporated.

The success of air pollution control plant depends, largely, on effective system design and layout [52]. Well-distributed flow is a key to the success of a plant. Since flow always follows the path of least resistance, designers must be extremely careful in arriving at correct duct diameter, duct velocity, duct routing, etc. The duct system must have minimum pressure loss; therefore, the number of unnecessary bends, contractions where high friction loss may occur, must be minimized. Dust accumulation in the duct should also be avoided, as in worse conditions it may cause self-ignition through static charge. Further, system design should be such that pressure or, in other words, flow is well balanced in branch and main ducts. Proper synchronization in all the above-mentioned units is very important for successful filtration operation.

4.2. Dust-laden air-feeding zone

The primary aim of the operation is feeding the dust-laden air in a continuous manner and maintaining the required feed rate. The dust-laden air from all process units is drafted through the draft fan and supplied to the system through the duct arrangement. The characteristic of flow in pipe depends on the velocity profile of the flow, e.g. for very low velocity, laminar flow does not exist; hence, flow in duct is considered to be turbulent. For a fluid flow in a duct of constant diameter, the average velocity is parallel to the centerline of the duct and the velocity at the duct wall is zero [52]. When a fluid flows through a pipe or duct, the amount of pressure can be expressed as the sum of static and velocity pressures. Static pressure (p_s) is a pressure that acts in all directions in a fluid confined in enclosure whether in motion or at rest. Static pressure is independent of velocity pressure. When static pressure is below atmospheric pressure, then it is said to be negative and vice versa. Velocity pressure (P_v) is generated due to the flow velocity. It is a type of kinetic energy, which is exerted along the flow and is always positive. This can be represented as

$$P_v = \rho v^2/2g,$$

where

P_v = Velocity pressure, mm WG,
ρ = Fluid density, kg/m^3,
v = Fluid velocity, m/s,
g = Acceleration due to gravity, m/s^2.

Design of the duct should be such that the total pressure is well balanced in all branches and the main duct. Capture devices are usually hoods or direct exhaust couplings attached to a process vessel. Direct exhaust couplings are less common, requiring sweep air to be drawn through the process vessel, and may not be feasible in some processes. In the design of the duct, the duct should be free of unnecessary bends and the system should be designed to have a balanced flow in ducts [52].

Different types of arrangements are available to feed the dust-laden air into the filter unit. Design of inlet has a great influence on the performance of fabric filter. There are many different inlet designs, such as tangential, involute, radial, high entry, etc. each having specific advantages for various applications. Each inlet design has its own advantages and disadvantages. For example, an involute scroll inlet puts dust into a cyclonic spin allowing heavier particles to fall into the hopper, thus eliminating the need for a cyclone pre-cleaner. However, this configuration is not suitable for abrasive or light and fluffy dusts [92]. Each inlet design will require different energy levels to move the process air though them. The differences may only be 12.5–25 mm of water column, but that pressure difference on a

200-horsepower fan can have a significant impact on its amperage draw. Therefore, inlet design should be the most efficient one regarding particulate collection and energy requirement, and not just the cheapest alternative [99]. As regards the entry point, top entry is more suitable for fine dust (such as talc) on applications with low-dust concentrations and with higher air-to-fabric ratios. Top entry generates a down flow of air throughout the collector, due to which it aids the movement of particles toward the hopper. In the case of coarse dust (such as sand etc.) and high-dust loads, hopper entry is preferred. Heavy particles naturally gravitate to the hopper. However, in most of the designs, inlet is provided at the hopper for smaller fabric filters (Figure 7). In this case, a counter current gas flow is usually referred to as *can velocity*, which is an important bag sizing criterion for online pulse-jet cleaning. The upward gas velocity inside the filtration chamber is termed as *can velocity*, which is the flow rate divided by the horizontal cross-sectional area of the baghouse less the cross-sectional area of the bags. In the case of the online cleaning process, upward flow of gas (typical flow pattern for bottom feed) should be limited to settling the dust particle at the bottom of the hopper. The can velocity should not exceed 11.8 to 16.5 cm/s, with the lower value more suitable for fine dust that does not settle easily.

Although, in the said design, heavier dust particles can settle before interacting with fabric filter, there is a risk of re-deposition of smaller size particulates back into the filtration process. However, this design is popular because of simplicity and cost-effectiveness. In the case of cassette/envelope-type filters, entry of dust-laden air/gas is avoided by providing an inlet at the top location (Figure 16). This is possible because the cassettes are removed from the side and not from the top. In the case of large filters with trough hopper, an inlet entry in hopper gives a poor gas distribution [81]. To avoid this, gravimetric inlets through introducing baffle plates along the length of the filter are quite effective (Figure 17a). Gravimetric inlets have two advantages: this design allows larger particles to settle down before they enter the filtration zone, whereas plenum design causes the gas to distribute in the whole filter. In the case of a parallel gas flow situation, gravimetric inlets with common outlet plenums are used, as shown in Figure 17b [13]. Furthermore, there could be a tangential inlet instead of straight inlet for dust feeding in the filter house.

Use of a diffuser/baffle plate at the entry of the filter unit can enhance the performance of filtration. Dust-laden air enters the collector through a diffuser, which absorbs the impact of the high-velocity dust particles, distributes the air, and reduces its velocity. Several diffuser designs have been proposed by machine manufacturers, such as impingement plates, perforated disks, perforated mailboxes, Cascadair, and Expandiffuse by Mikropul [122] etc. There are also various patented methods [123,124], in which some unique designs of gas-inlet geometry for feeding dusty gas in pulse-jet baghouses can be found. Figure 18 shows perforated mailbox and Cascadair arrangement for feeding the aerosols in the filter unit. While designing the diffuser, it is very important to study the flow pattern of aerosols inside the filter unit to avoid dust re-entrainment and uneven dust discharge.

4.3. Filter unit

In a filter unit, selection of filter media is one of the most important criteria. As with other separation processes, filter media are a vital part of baghouses. Although they normally account for a relatively small part of the total cost of the collector system, they may be one of the most critical factors governing the success of the system. Filter media should provide high-filtration efficiency and low pressure drop and should have longer life. High-filtration efficiency indicates less penetration of particles and hence less emission. Emission is dependent on a vast number of factors and increases with the following:

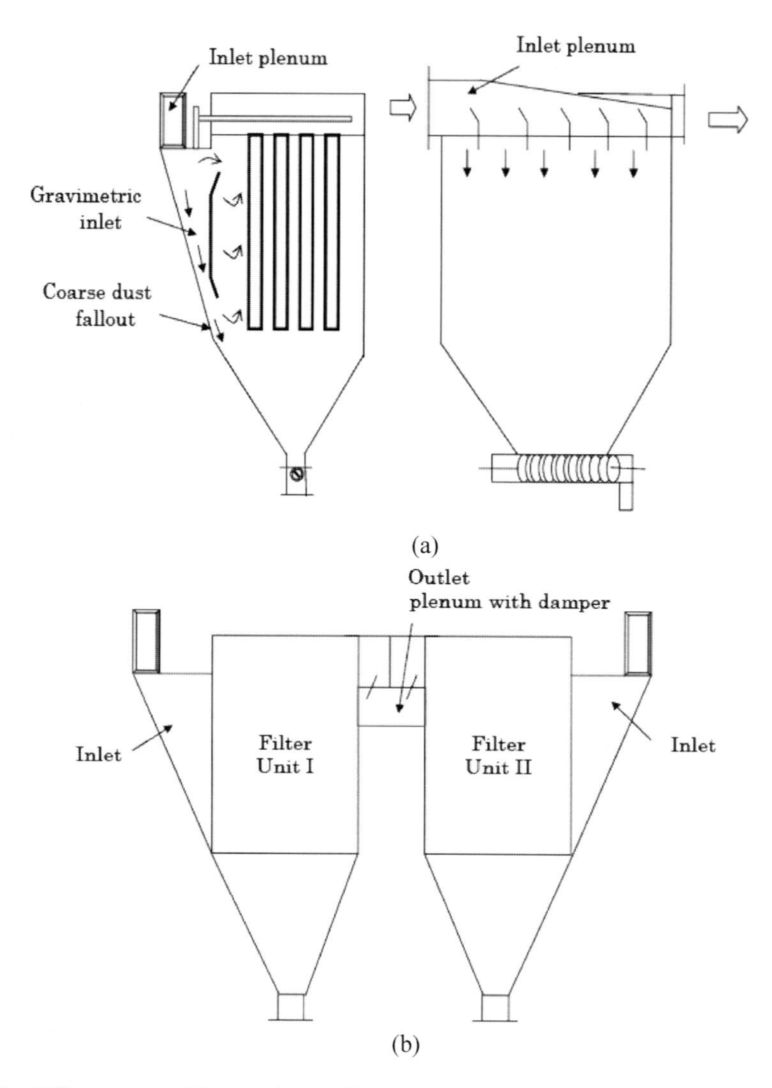

Figure 17. Different types of dust entries. (a) Gravimetric inlet; (b) Inlet with common outlet plenum.

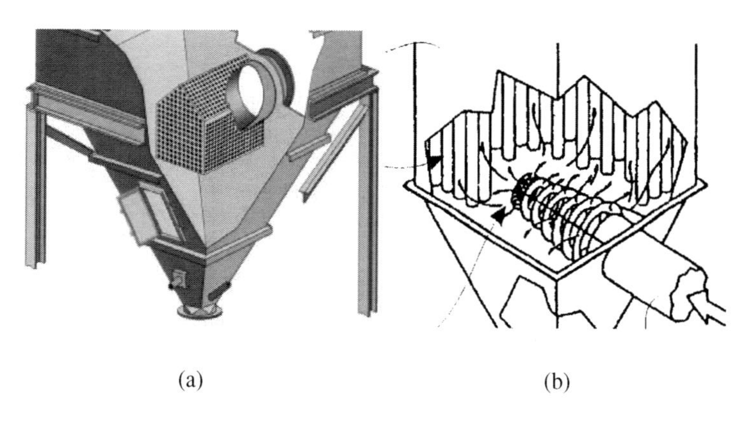

Figure 18. (a) Perforated mailbox; (b) Cascadair bottom inlet geometry.

- Inlet dust concentration but not proportionately;
- Decrease in particle size. Behavior is complex for particles ranging between 0.1 μm and few microns;
- Larger equivalent pore size of the fabric;
- Lower area density of the fabric;
- Increased filtering velocity;
- Increased cleaning intensity, i.e. increased pulse overpressure;
- Increased pulse frequency.

Designing of a filter unit including media should be appropriate in accordance with the nature of the aerosol. Pressure drop (Δp), is another very important baghouse design variable, which describes the resistance to airflow across the baghouse: the higher the pressure drop, the higher the resistance to airflow. During filtration tests, the pressure drop is measured by some form of differential pressure device with probes upstream and downstream of the media and is usually expressed in millimeters of Hg or inches of water. Conventionally, a U tube manometer is used. Modern tests use electronic processors and digital readout capability. During filtration, pressure drop is monitored to evaluate the performance of the filtration system. Cross et al. [125] developed a monitoring system for the mass and tension of bag filters. The pressure drop could be derived from the data obtained from such a system designed for full-scale bag filters.

While designing the filter fabric, it is often very difficult to meet the two contradictory requirements – highest level of filtration efficiency and minimum level of pressure drop. Improving the filtration efficiency of a fabric filter (through structural modification/higher material consolidation) leads to higher pressure drop for virgin filters. Most often for meeting stringent particulate emission, higher filtration efficiency is chosen even at the cost of higher initial pressure drop. During operation, the pressure drop originates from the drag forces offered by dust deposited outside and inside the filter media. Fine particles tend to penetrate into the pores of the filter media on the conditioning step and plug the filter media or slip out in the clean gas stream. The conditioned filter element has a permanent dust layer, which plays a role in additional filtration function. The resistance coefficient of the permanent layer is usually more than 10 times the value of the fresh bag filter element [126]. The total system pressure drop can be related to the size of the fan that would be necessary to either push or pull the exhaust gas through the baghouse. A baghouse with a high-pressure drop would need more energy or possibly a larger fan to move the exhaust gas through the baghouse [13]. For evaluation of the operational property of a filter unit, it is also important to determine the time required by the unit to reach the limiting pressure drop (say 1500 pa). For shorter duration of filtration cycle, the fabric is to be cleaned often, which means that the unit will consume more energy, and the dust emission in the cleaned gas stream will increase continuously [127].

In the process of filtration, pressure drop pattern is quite different for steady and unsteady processes (Figure 19). Due to the clogging of the media, pressure drop across the system gradually increases and, even at the steady state, quite a small and imperceptible change takes place in the normal operational stage. The residual pressure drop (the pressure drop across the filter just after cleaning) exceeds the pressure drop of the virgin filter element because of a residual dust layer, the occurrence of patchy cleaning, and/or the particles that have penetrated the filter medium and have not been removed by the cleaning action. This residual pressure drop and, henceforth, the total system pressure drop will continue to increase from cycle to cycle. In normal cases, it will reach a relative stable value after the so-called *conditioning* period. It can be seen in Figure 19 that a partially concave shape of

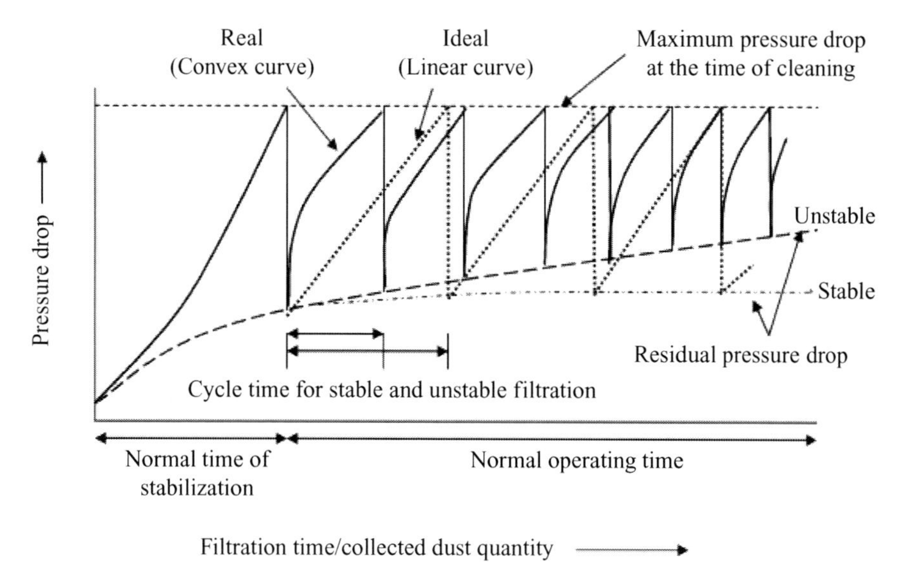

Figure 19. Fabric filter behavior in relation to pressure drop vs. time.

the pressure drop curve results during the first conditioning phase. This indicates particle separation on the inside of the filter (depth filtration) or, in the case of larger dust surface weights, a compression of the dust cake. However, a convex curve progression shows just after a few cleaning processes (cleaning cycles), which is determined by the structure of the cleaned off-filter surface. With only partial removal of the filter cake (patchy cleaning) or the formation in fissures, this effect may be very prominent and lead to a significant reduction in the filtration cycles [128]. A convex curve is also apparent in the case of the so-called stable operation.

In practice, the extent to which these effects occur especially depend on all relevant parameters influencing the operating behavior of a cleanable filter, especially the properties of the filter medium, the gas and dust properties (composition, temperature, chemical reactions, etc.), the system operation (filter face velocity, pressure drop flange to flange, etc.), the effectiveness of the cleaning system, and the geometric properties of the filter system (e.g. flow distribution) [128]. However, a stable operation condition will never be reached for some *active dusts*. In the case of filtering sticky or sintered dust, which is difficult to dislodge, unacceptable pressure drop will be reached quicker than normal in the lifespan of a fabric. In addition, when the fabric material is wrongly selected or filtration velocity is very high or cleaning is insufficient, or the fabric has got clogged due to condensation, pressure drop will tend to be higher even before the filter is stabilized.

In a steady state, dust penetration, as well as pressure drop, tend to be constant. For steady state, highest pressure drop is reached after a long operating time after which bags must be replaced. The rate of increase of residual pressure drop during the filtration cycle is very important because it determines the lifetime of the filter bag. A medium is said to be blinded when cleaning fails to remove residual solids that are adhering or are embedded to it, so that its resistance to flow remains unacceptably high [129]. Apart from fabric selection, a successful pulse cleaning of the filter is clearly essential to reach a low and stable conditioned pressure drop and to operate the filter over long periods. However, performance of an inappropriate filter element cannot be improved by increasing pulse frequency and pressure as both will lead to mechanical damages to the filter bag leading to shorter bag life. While designing the filter media, the knowledge of the effect of particle properties (shape, size, roughness, adhesion, cohesion characteristics, etc.) and operating variables on fabric filtration is required [130–132].

Apart from clogging of fabric pores by particles and problems related to dislodging of particles, durability of the filter media also depends on mechanical, chemical, and thermal degradation. Mechanical degradation is mostly due to repeated flexing of fabric with dust particles trapped inside, and stressing of pores during pulse cleaning. The cleaning strength and frequency contribute to the flex abrasion. Furthermore, improper slack between the filter element and the supporting cage leads to quicker flex abrasion [13,133]. Repeated flexing over a very large number of cycles also leads to fatigue. Particles of sharp edges create pin holes, which subsequently grow into vertical streaks known as plastic deformation or cutting wear. The problem is further aggravated with the increase in filtration velocity and dust concentration as they are directly related to the amount of trapped particles inside the fabric structure. Chemical degradation occurs due to moisture, temperature, acids, alkalies, and oxidizing agents, whereas thermal degradation mainly depends upon particle and gas temperature. Degradation of material accentuate in combination with thermal and chemical environment.

The factors which influence the life of a filter can be summarized as follows:

- The speed at which the particles hit the fabric decides the filter life. Velocity more than 2 m/s is undesirable;
- Dust concentration/the total dust hitting the fabric;
- Adhesive properties of dust;
- The geometry and hardness of the particles;
- The shape and size of the particles. Smaller particles lead to clogging, whereas large particles cause another kind of wear, which is known as 'fiber-picking';
- Chemical nature of the particles/aerosol;
- Thermal environment/thermal characteristics of particles;
- Presence of moisture in the air;
- Explosive characteristics of particles;
- Cleaning cycles and pressure;
- The physical, mechanical, chemical, and thermal characteristics of fabric.

There is an empirical formula to calculate the life of bag filters based on the change in the average face velocity, as given below [13]:

$$L_b = 4.0\,(0.61\,/V_b)^{0.6},$$

where

L_b = life of bags in years,
V_b = average face velocity in m/min.

The above formula is based on the physical degradation of the fabric considering increased particle clogging inside the fabric structure at higher filtration velocity. With baghouse filters, once the application details have been taken into account, the environment in which the filtration will take place must be closely examined before determining the type of the filter bag to be selected. The following factors are, in general, important for selecting the type of fabric filter [13,90]:

1. *Particulate emission*: As the emission norms become stringent, it is highly desired that the filter fabric is capable of capturing particulates even below 2.5 μm.
2. *Temperature*: Operational temperature is probably the most important, irrespective of the gas conditions. Filtration applications are subdivided into two groups: low-temperature filtration, which operates at local ambient temperatures, such as milling, grinding, mixing, and bagging operations; and high-temperature filtration, where heat is induced or is a part of the process, as with utility boilers, incinerators, dryers, or kilns.
3. *Adverse gas and chemical conditions*: The ability of the fabric to resist degradation from expected levels of acids, alkalis, solvents, and oxidizing agents is important to be assessed at the operational temperatures of the filtration application.
4. *Humidity*: Resistance to hydrolysis at expected temperature and humidity levels should be ascertained.
5. *Dimensional stability*: The filter media should resist shrinkage and stretching at the expected temperatures and gas conditions.
6. *Type of filter unit*: The amount and duration of pulse pressure decides the stress level in the fabric; and hence the type of the fabric should be suitable under the given conditions.
7. *Durability*: The filter fabric should be able to sustain damage under chemical and thermal environment, and should be resistant toward internal abrasion and flex fatigue. It should also have less tendency toward clogging for a longer life.

The selection and designing of a filter element is discussed in Section 5. In the designing of bag size, the following parameters are considered:

- The volumetric flow rate of the air stream;
- The operating temperature and gas property;
- Concentration of dust particles;
- Air-to-cloth ratio;
- Can velocity.

Volume of air, concentration of dust, temperature of gas stream, dew point, nature of particles, etc. can vary widely depending on the nature of the industry and even at the different process steps in the same industry. Table 7 shows dust concentration, air volume, and particle size distribution at various stages in a modern cement plant [52,134]. In the cement-making process, dust is generated at each stage of production. In operations like raw mill grinding, coal grinding, and clinker grinding, high dust generation to the tune of 650 gm/Nm^3 is experienced. Besides this, the dust is usually very fine (50–80% is less than 5 μm). Further gas temperature in clinker cooler usually ranges from 200–250°C. All these factors should be taken into account while selecting the dust collectors. Considering various aspects, the detail about the size calculation of the bag (i.e. number of bags) is given in the appendix. Number of bags can range from a few bags to large numbers (e.g. 4000) depending on the gas volume.

In most of the cases, the structure of the filter element for all types of filtration is in the form of circular cross-section. The filter fabric is stitched to form the desired dimension. In pulse-jet filter design, bag fixing is entirely different in contrast with reverse jet and mechanical system. To support the bags during pulsing, and to avoid collapsing, cages are introduced inside the bags (Figure 20). Cages are mounted over the top frame without causing stress to the bags. There are many ways to fix the bags over the top bag mounting frame [94]; bags are held firmly in place at the top and usually have an enclosed bottom (the

Table 7. Dust generation, particle size distribution, and air/gas volume in different sections of a modern cement plant.

Section	Vent (air/gas) volume	Range of dust generated (gm/Nm^3)	Particle size distribution Size (μm)						
			0–5	5–10	10–20	20–30	30–40	40–60	>60
Crusher	50 $m^3/min/m^2$ of feed hood opening	5–15	15.32	11.08	13.08	21.00	19.80	11.00	8.00
Raw mill	1.5–2.5 Nm^3/kg	650	61.00	6.00	6.00	3.00	2.00	3.00	19.00
Rotary kiln	1.7–2.0 Nm^3/kg	50–75	76.00	5.00	4.00	2.00	1.00	1.50	9.50
Clinker cooler	2.2–2.6 Nm^3/kg	5–10	6.00	2.00	2.00	1.50	1.00	1.50	86.00
Cement mill	4 times mill volume/min	60–100	50.00	7.50	5.50	3.50	3.00	4.00	25.50
Packing plant	35 m^3/min per filling spout	20–30	65.00	4.00	4.00	1.00	1.00	2.00	22.00

bag is sewn closed at the bottom). In one of the designs, 'O' rings are sewn at the top cuff of the bag. A properly designed O-ring groove with proper fitting doesn't allow the passage of pressurized gases. O-rings are made of elastomeric materials such as rubbers and plastics. Flexibility while being deformed under pressure is their greatest sealing advantage. They deform under pressure and then return back to their original shape once the pressure is removed. In one of the designs, the annular flange is used, which lays flat on the bag-hanging plate. In another design, a thin stainless steel band, called snap band, is inserted on the top cuff along with a double-beaded gasket stitched into the cuff. For easy installation and changeover of bag and cage, Mikropul [122] offers a new design incorporating an elastomeric resilient collar to be used through pop-top technique in any application where snap-in bags are used. Figure 21 shows arrangement for snap bands and Mikropul resilient collars, respectively. In one of the preferred embodiments by Donaldson, it is possible to have quick and easy release of filter bag cages by applying low-level load ($\approx$ 27 N) on a handle fitted with the cage [92]. The system is especially suitable for applications like cement where bags tend to stick to their cages.

The bag is sealed after the cage and venturi are bolted in place. If the bag designs are envelopes or cassettes, they are fixed with vertical fixing plates to keep them horizontal, and cleaning air is introduced from the side. Bags are arranged in an array: the minimum distance between the two bags is kept as 50 mm to avoid abrasion. On the pulse valve side, distance between the bags is fixed depending on the valve size [13].

As regards the use of cages, different types of cage design are shown in Figure 22. Envelope bags require more elaborate interior support than cylindrical bags. Dimensions of the cage depend on the dimensions of the bag and the bag slack. The clearance between the bag and cage is termed as bag slack and is determined by the ratio of the circumference of bag to the cage. Usually, slackness ratio of the bag is 1.05. Bag slack is necessary to achieve filtration performance during pulsing and to avoid the damage to the bags by the cage friction. In the construction of the cage, the number of wires may vary from 8 to 40 depending on the fabric and the application. For glass felts, the number of wires is around 20. Fewer wires provide better cleaning due to flexing of bags. In general, 8–10 wired cages are used [13]. It is important to note that filter media sticking to the cage wires interrupt necessary filter bag flex during pulse cleaning. The filter media may stick to the cage wires due to reactive degradation with chemical changes in the galvanized finish, or more physical

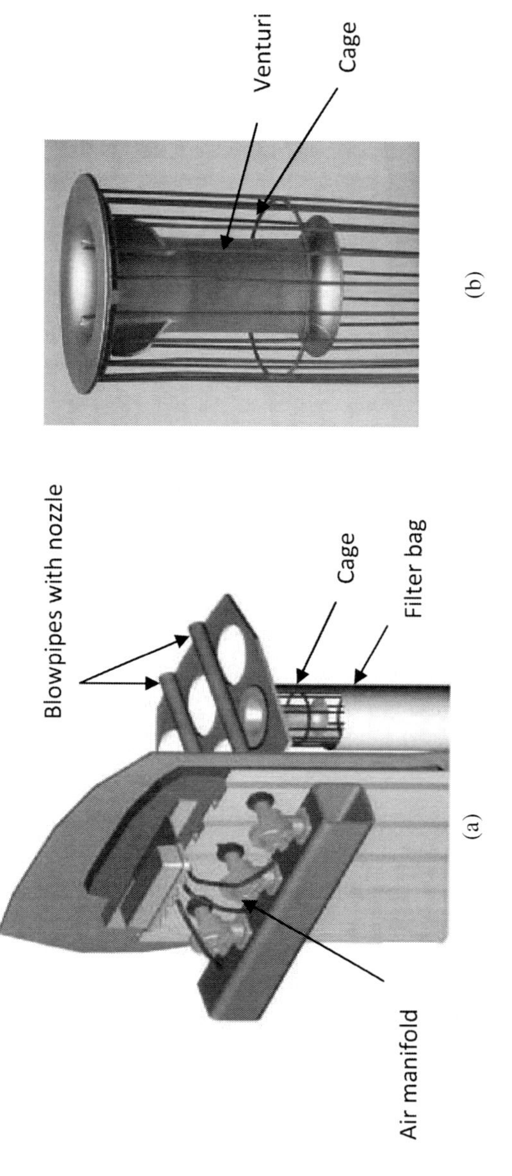

Figure 20. (a) Bag mounting arrangement; (b) Arrangement of venturi.

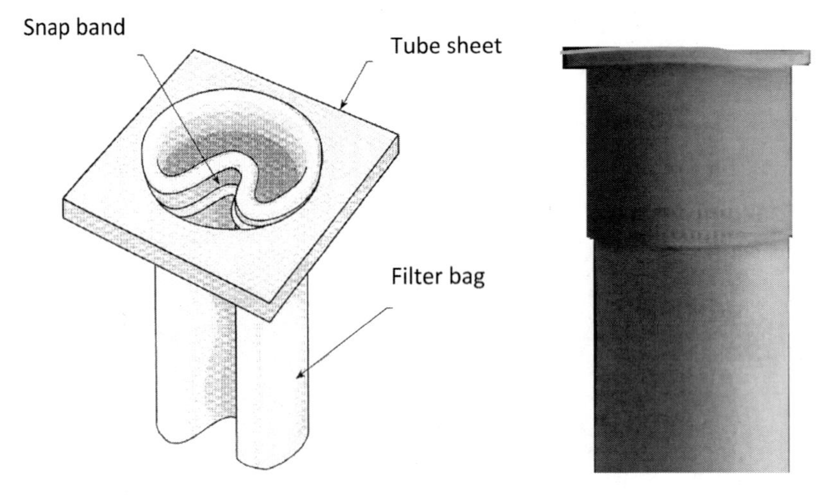

Figure 21. (a) Snap band with filter bag; (b) Bag with an elastomeric collar.

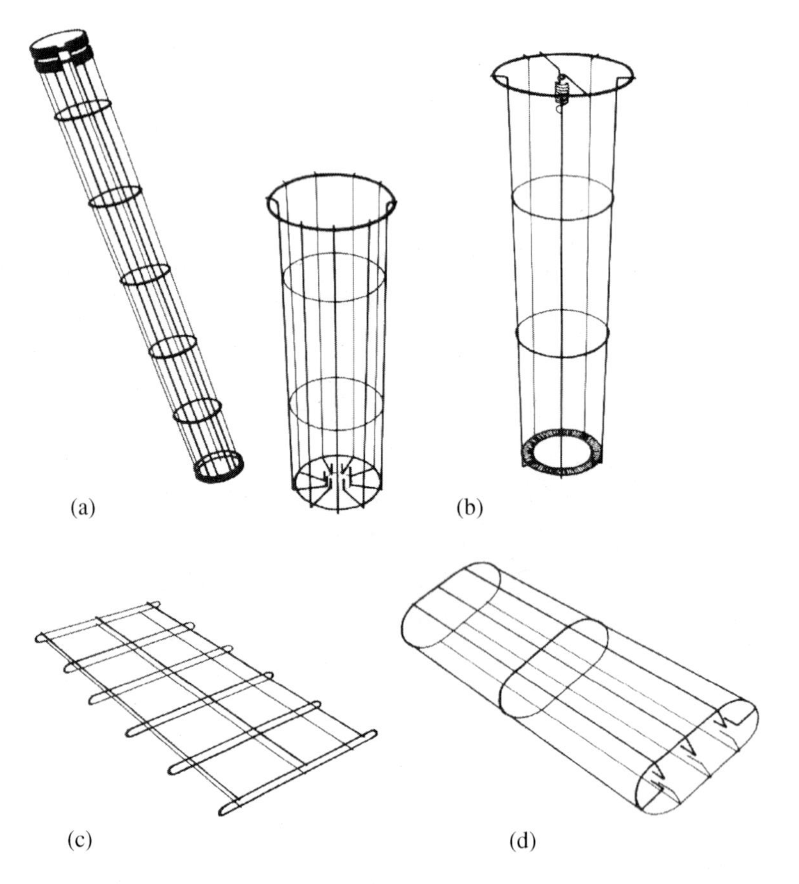

Figure 22. Different types of cages. (a) For cylindrical bags; (b) For double-wall cylindrical bags; (c) For envelope bags; (d) For oval bags.

adhesion with the severely corroded cage wires. Therefore, material composition and finish of the cage play an important role.

In all these design outlays, the most important factor in baghouse is regeneration of bags through pulse-jet cleaning after each filtration cycle. A pulse-jet collector can be operated to get the maximum output – higher efficiency, lower power consumption, and a much longer filter element life [135]. Due to this cleaning action, the bags undertake a repeated number of filtration cycles with lesser pressure drop. The intelligent pulse-cleaning control technology offers direct connection to most PLCs with user-specified field bus communications, such as DeviceNet, Ethernet, Modbus, and Profibus. However, basic pulse-jet cleaning systems consist of the following [13,52,136,137]:

1. *Air manifold*: Air manifold holds the compressed air to a desired pressure and quantity. The manifold stores the pre-determined quantity of air in the pressure range of 200–700 kPa, which is to be pulsed for filtration purposes [13,52].
2. *Filter regulator unit*: The function of filter regulator is to supply clean air to the manifold for the pulsing action.
3. *Pulse valve/other systems*: There could be a wide range of valves, such as diaphragm pulse valves, piston pulse valves, fully immersed piston valves, etc. and several other systems are available for cleaning purposes [13,21,136,138–141]:

    - Diaphragm pulse valves operated through solenoid consist of one main diaphragm and a pilot diaphragm. In a closed condition, air pressure is balanced between two diaphragms. Once the solenoid is energized, the air between the diaphragms escapes through the pilot port. This causes imbalance of pressure, which enables the air to rush into the blowpipe and to the bags. The entire process takes place in milliseconds. After completion of this process, diaphragms go back to their respective positions and pressure is restored until the next operation of the valve [13].
    - To increase the flow and to make the system more compact, the diaphragm valves were sometimes fixed directly to the tank (in general, the valves are typically piped onto the pressure tank). A new piston-style valve proposes to offer high filter-cleaning efficiency. These valves are fully immersed in the power pulse tank system, which result in effective air blast for optimal cleaning effect and low energy consumption [89].
    - In a commercial low-pressure high-volume pulse-jet unit, the cleaning system is applied only to baghouse filters and the most frequently housed cylindrical body using a rotating manifold to distribute the cleaning air. The cleaning system utilizes a positive displacement blower, a reservoir, and a solenoid-actuated diaphragm valve to deliver short bursts of air to the filter bag. The positive displacement blower compresses air into the reservoir until it reaches a specific pressure (55–100 kPa). At that point, a timer assembly energizes a solenoid valve, which in turn allows a large diaphragm valve to open between reservoir and cleaning manifold [21].
    - A method consisting of an apparatus for improved pulse-jet cleaning of industrial filters is proposed in which rotation of one pipe relative to another pipe about a shared longitudinal axis causes apertures in the pipes to align intermittently. When the apertures are aligned, pressurized air is fed through a pulse valve into the inner tube of the two and flows out of the nested pulse-pipe arrangement in a short, energetic pulse (Figure 23). The intermittently injecting pulse is characterized as low-pressure, high-volume flow of gas for cleaning the bag [137].

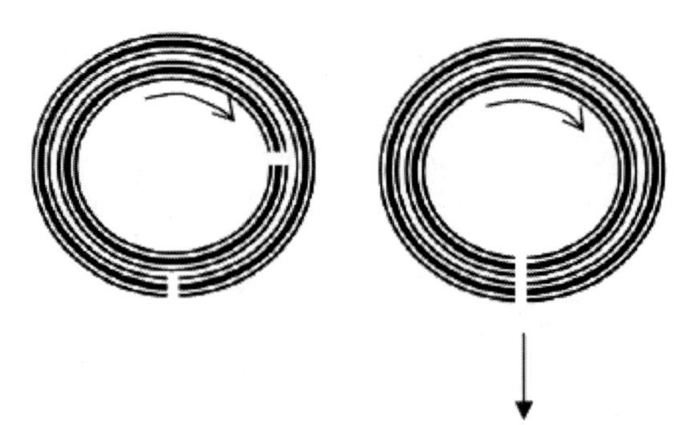

Figure 23. Principle of pulse-jet injection.

- In a commercial system, a specially designed mechanical valve is placed inside the top of the filter to provide low pressure (0.7–0.9 bar) but high-volume cleaning pulse to clean up to 11-m long bags. A low-pressure compressor providing the necessary cleaning air to the receiver is placed directly below the Pulse Air Distributor (PAD) valve. The PAD secures that all bags are cleaned one by one exactly with the optimum cleaning energy. A PLC control makes it possible to optimize pulse time, cleaning energy per bag, etc. and the normal procedure is automatically to start and stop cleaning sequences to keep the pressure drop across the filter within the pre-set limits. This unique individual bag-cleaning principle will, in all normal operating conditions, result in very long cleaning intervals, which together with the low-pressure, high-energy pulse provides a very gentle cleaning and long bag life [139].

4. *Blowpipes*: These are used in high-pressure, low-volume and medium-pressure, medium-volume pulse-jet units. These pipes are connected to the outlet of the valves and span over the bags aligned in rows. Each pipe has nozzles depending on the number of bags to be cleaned in that row.
5. *Venturi/diffuser element*: Venturis are installed in the neck of each bag or may be fitted at the top of the tube sheet, below the nozzles of the pulse tube. With the venturi located in the filter bag, a negative air pressure is generated during bag cleaning. Venturi above the tube sheet eliminates this problem [140]. The purpose of venturi is to increase the intensity of the pulse pressure injected into the bag from the nozzle [142]. The optimum design of venturis has been described in the literature. Rothwell et al. [143] investigated the use of a venturi and concluded that a venturi with a large throat develops lower pressures but induces a higher mass flow compared to a venturi with a small throat, which develops higher pressures but induces a lower mass flow. The higher the cleaning pressure, the larger the optimal throat diameter for effective cleaning [143].

In the case of a ceramic filter, Choi et al. [117] summarized the basic criteria of venuri design as mentioned below:

- Entrance to the throat should have an included angle of 19–23°. Sharp edge transitions should be avoided.
- Length of the throat should be the same as the diameter.
- Included angle of the outlet cone should be 15° maximum.
- Exit cone diameter should be the same as the entrance cone diameter.
- Ratio of the entrance cone diameter to the throat diameter should be at least two.

A diffuser element was also used to regulate cleaning pressure on the bag. This is simply a cylindrical sleeve made of perforated stainless steel that fits inside the cage of a pulse-jet and extends up to one-third to two-thirds down the length of the bag. It has been shown to reduce bag-cleaning deceleration at the top and increase it at the bottom. Excessive fabric flexing at the top can reduce the life of the filter bag [142,144]. However, diffuser elements are no longer used in practice.

It is very important to make the baghouse system energy efficient. There could be around 85% of the energy consumption due to the pressure loss during filtration, where the deposited cake can have the major role, and the other 15% of energy is needed for the cleaning process (online operation, pressurized air of 0.6 MPa) [145]. Through proper design of the filter unit, as well as setting of operating parameters, the system can be made energy efficient. The filter unit is enclosed inside a casing, which is, in general, designed with mild steel or stainless steel. This design is normal to all types of filtration units unless the filtration deals with high temperature and pressure. These panels should be capable of withstanding the negative/positive pressure generated during the filtration process. In industries, the selection of casing materials is based on the all-possible combinations of loads, such as live load, dead load, wind/earth quake, temperature, pressure, etc. [13].

4.4. Dust disposal system

In the cleaning process after each filtration cycle, the dislodged dust is removed with the help of a hopper, rotary air lock, and screw conveyor. The hopper is characterized by the casing size, valley angle, and the rotary air lock size. Various designs of hoppers are available, like trough hopper, pyramid hopper, which can fit with the main casing and to the rotary airlock. Rotary air lock prevents the air leakage from the chamber and discharges the dust at a specified rate. Air leakage increases the load on the filter. The design and speed of screw conveyor or rotary airlock depend on the rate of dust fallout from the baghouse to hopper [52].

4.5. Fan (air mover) and discharge stack

In any air pollution control system, air/gas has to move from the point of off-take to the control equipment by the action of fan. The fan has to overcome the air/gas drag by the filter fabric, duct, and resistance offered by several other passages. Pressure drop in the system should be accurately determined for selecting the correct fan. Fans may be classified according to the direction of airflow through the impeller. Two most common types of fan used in industries are axial flow and centrifugal flow fans. However, cross flow and mixed flow fans are also possible [21,52] From the fan, the air flows out through the discharge stack.

4.6. Auxiliary attachment

In case dust collectors are used in a plant to control indoor air quality, keep equipment clean and/or recover high-value process dusts, as many plants consider re-circulating the air back into the plant downstream of collector instead of exhausting it outdoors. A suitable auxiliary system is very useful for energy conservation. However, when using re-circulating dust collection systems, special safety and performance concerns must be addressed [146]. A pulse-jet cleaning system [147] can be augmented with sonic cleaning method. Sonic energy is introduced into the baghouse by air-powdered horns, and the shockwaves produced generate forces that tend to separate dust from the bags and the interior surfaces of the filter unit. The horns typically operate in the range of 125 to 550 Hz (more frequently in the range of 125 to 160 Hz) and produce sound pressures of 120 to 145 db. When properly applied, sonic energy can considerably reduce the mass of the dust on bags, but may also lead to increase dust penetration through the fabric. Increased penetration reduces the efficiency of the baghouse [107].

Some typical auxiliary equipment, such as spray chambers, mechanical collectors, and dilution airports, may be needed to either pre-condition the gas/or enhance particle capture before it reaches the fabric filter. Spray chambers and dilution airports decrease the temperature of the pollutant stream to protect the filter fabric from excessive temperatures (Figure 24). When a substantial portion of the pollutant loading consists of relatively large particles (more than about 20 μm), mechanical collectors, such as cyclones, are used to reduce the load on the fabric filter [107].

The combined use of fabric filters with sorbent injection systems has been utilized for many years in the municipal incinerator as well as other industries as a way to enhance the removal of inorganic gases (SO_2, N_XO_Y, etc.), organic gases (dioxins, furans, etc.), Hg, and a wide range of heavy metals. When the sorbent is injected into the flue gas, it mixes with the gas and flows downstream and, in the process, gaseous matter is removed. This also provides an opportunity for the Hg in the gas to contact the sorbent where it is removed, which is known as *in-flight capture*. The sorbent is then collected in the particulate control device where there is a second opportunity for sorbents to be exposed to Hg present in the gas [148].

Cleaning processes of flue gases from waste incinerators consist of arrangements of equipment like electrostatic precipitators, fabric filters, wet scrubbers, spray absorbers, entrained flow absorbers, etc. High removal efficiency is necessary to comply with the current emission limits for Hg in flue gases to be removed from waste incinerators. Each type of equipment has its typical range of removal efficiency for Hg and its compounds. A summary of the various ranges of Hg-removal efficiencies by various technologies is given in Table 8. It is quite apparent that spray absorbers and other types of gas–solid contactors, both with special absorbents added to the feed and followed by a fabric filter

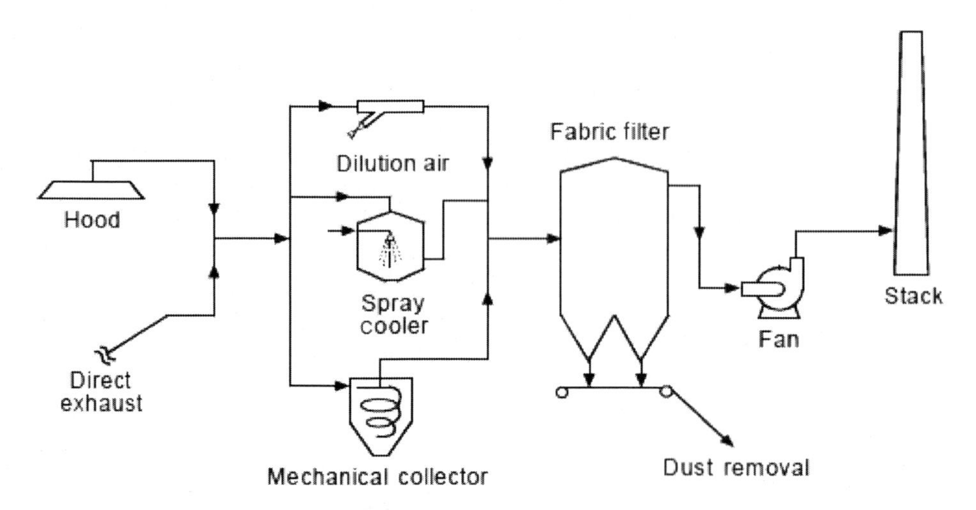

Figure 24. Typical auxiliary equipment used with fabric filter control system [107].

Table 8. Mercury removal efficiencies in industrial equipment [15]. (Reprinted from D.V. Velzen, H. Lagenkamp, and G. Herb, *Review: Mercury in waste incineration, waste management and research*, 20 (2002), pp. 556–568, with permission of Sage Publication Ltd., UK.)

Equipment	Temperature (°C)	$HgCl_2$ (%)	Hg(O) (%)	Overall (%)
Electrostatic precipitators	180	0–10	0–4	0–8
Wet scrubbers	65–70	70–80	0–10	55–65
Wet scrubber with conditioning agent	—	90–95	20–30	76–82
Spary absorbers + fabric filter (limestone)	130	50–60	30–35	44–52
Spary absorbers + fabric filter (special absorbant added)	—	90–95	80–90	87–94
Entrained flow absorbers + fabric filter (special absorbant added)	130	90–95	80–90	87–9
Circulating fluidized bed + fabric filter (special absorbant added)	130	90–99	80–95	87.98

(not an ESP), are efficient tools for the removal of Hg. A simplified flow scheme of the flue gas purification of dry process plant is shown in Figure 25. Inside the reactor, finely ground hydrated lime is mixed with the flue gas. SO_2 and HCl react with the lime and form $CaSO_3$ and $CaCl_2$. The flue gas and the lime exiting the reactor are cleaned in the fabric filter. The amount of carbon added is 0.7% of the lime addition rate. The system is reported to operate very well and, typically, a Hg outlet concentration of less than 20 $\mu g/Nm^3$ is obtained, which is well below the limit of 50 $\mu g/Nm^3$ [15].

4.7. Safety features for explosive and/or inflammable dust

In order to describe the explosion risk posed by dust, a number of factors are needed to be described. These include particle size, explosion limits, the maximum explosion pressure, the destructive power of the combustion, moisture content, and the minimum ignition energy required. Furthermore, an examination of industrial processes is required on the basis of possible ignition sources, explosive volumes, operating temperatures, and an assessment of the possibility of a dust explosion under given conditions. In case of combustible dust, the design and operation of the dust collector must meet minimum safety standards. Recommended practice on static electricity also has a significant impact on baghouse and dust collection system design. There are several guidelines provided by the National Fire Protection Association (NFPA), *Atmosphères Explosibles* (ATEX) Directive, etc. [149,150] for avoidance of fire hazards. Recently, the Occupational Safety and Health Administration (OSHA – main federal agency charged with the enforcement of safety and health legislation) issued a Combustible Dust National Emphasis Program, directive number CPL 03-00-006, effective from October 18, 2007, which was re-issued as directive number CPL 03-00-008, effective from March 11, 2008. This document contains policies and procedures for the inspection of workplaces that create or handle combustible dusts, which include but are not limited to, metal dust, wood dust, coal and other carbon dust, plastic dust, and organic dust such as sugar, flour, paper, and soap. The directive recommends that NFPA explosion protection standards be consulted for the design requirements associated with the explosion protection of the process equipment, including dust collectors [151,152].

NFPA has a wide variety of codes that are used by regulatory agencies to determine compliance with industrial safety standards. Design options for the dust collectors should

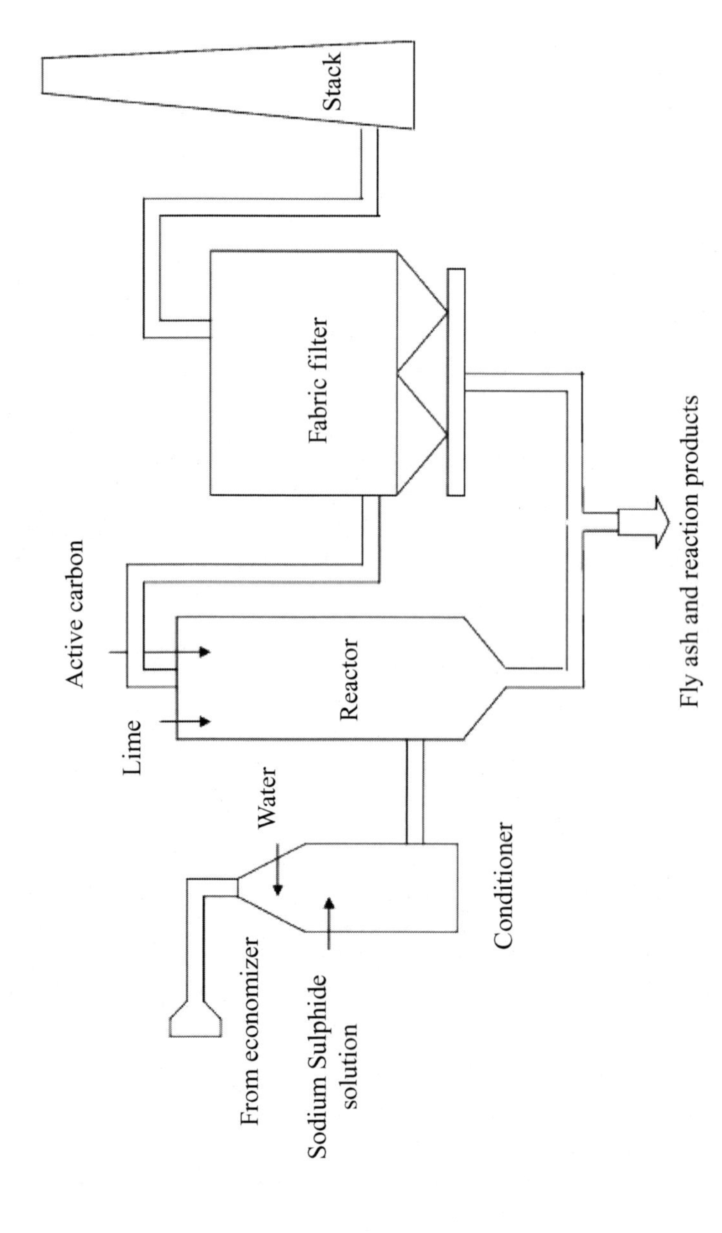

Figure 25. Scheme of a dry process. [15]. (Reprinted from D.V. Velzen, H. Langenkamp, and G. Herb, *Review: Mercury in waste incineration, waste management and research*, 20 (2002), pp. 556–568, with permission of Sage Publication Ltd., UK.)

include considerations for the combustible nature of the dust, the mode of operation for the process system, and/or explosion suppression systems. The ATEX directive defines the type of protection provided by enclosures, based on limiting the maximum surface temperature of the enclosure and using dust-tight and protected enclosures to prevent dust entry. There is a report on the designing of a dust collector system and equipment following the above directives [153]. While constructing the filter unit, proper selection and design of filter media plays a vital role. Various constructional features of filter media for handling explosive and inflammable dust particles are discussed in a separate section.

Some of the safety features of filter unit are mentioned below [13,52,149,150]:

- The most cost-effective solution for controlling dust usually begins at the source. Improved designs in transfer chutes and transition ducts can reduce the amount of dust generated during material-handling operations and reduce the amount of collection devices. For all dry collectors, care is taken by minimizing flat surfaces so that dust accumulation is avoided. All leakages are to be prevented to avoid fresh air entry into the circuit.
- Immediately before the explosion, pressure inside the vessel increases very rapidly and reaches a particular value that must be released. Explosion vent is a device that allows the release of explosion pressure into the atmosphere through an opening with limited or no damage to the main equipment (in this case, dust collector). In the case of an explosion, vents will rupture/release and the explosion pressure is vented.
- Proper earthing of the control equipment ensures that any static electricity generated is instantaneously discharged. In the case of a fabric filter, it is to be ensured that the casing, bags/bag-holding plates are grounded. It is the bags that will accumulate charge and the same is to be discharged effectively. Therefore, a good contact has to be there between bags and holding plate. Further, antistatic fabric ensures proper discharge of static electricity.
- Bag cleaning should be more than normal so that dust is not allowed to build up on the bags. Bag cleaning should be continued well after the stoppage of the plant to ensure complete dust removal from the bags.
- Hopper must be continuously evacuated during the running of the filter and complete evacuation is also to be ensured after the stoppage. Dust discharge devices, such as rotary air lock, screw conveyor, must function well and continue running even after the stoppage of inlet feeding. A hopper-level indicator is to be provided to prevent dust accumulation in the hopper with trip set point. A zero-speed monitor should be provided to ensure that rotary air lock and screw conveyor drive are working.
- The gas temperature is to be maintained well above the dew point temperature to avoid condensation in case of handling coal dust. Condensation helps the deposition of coal and enhances smouldering.
- The introduction of inert dust/gas may reduce the combustibility of dust by absorbing heat. Inert dust also reduces the rate of pressure rise and increases the minimum concentration of combustible dust necessary for ignition. However, some inert dust, such as silica, increases the dispersibility of combustible dust, which is harmful and should not be used. Inert gas (CO_2 and N_2, as per availability) is also used; however, CO_2 system is more common and cheaper.
- Inerting systems should be automatically started upon notification of high-temperature alarm conditions. It is necessary that the concentration of oxygen is limited to a set value. Therefore, *Dosing* is to be given at all critical points by a

piping network at the outlet, bag filter inlet, filter inside, etc. It is to be ensured that sufficient quantity is injected to dilute the oxygen level. In the case of coal mills, lime-dosing systems can ensure precaution against smouldering. The dosing is to be installed in the inlet ducting system between the coal mill and the filter.

- The operating temperature should be monitored, especially in flammable dust applications for early signs of combustion. CO and O_2 analyzers should also be provided before the bag filter inlet with alarm and trip set point so that the collector is isolated immediately if there is a possibility of explosion. In the event of a suspected fire, the dust collector fan should be stopped and the system should be isolated. The cleaning cycle must be stopped so that additional dust is not put into the suspension.
- Care should be taken so that dust is continuously evacuated from the hopper. Quick auto shut-off dampers should be provided at the inlet and outlet of bag filter to ensure complete isolation of the filter.

5. Selection and designing of filter media

5.1. General consideration

Worldwide needle-punched nonwovens are being used predominantly for filter bags. Needle punch filter media permits higher filtration efficiency at relatively lower pressure drop. Other than needle-punched fabric, spun-bonded, chemical-bonded, and hydroentangled fabrics are used. Manufacturing of nonwoven filter through three basic processes, such as needle felting, spun-bonding, and spun-lacing, is schematically represented in Figure 26. Although there is some growing interest on spun-bonded and hydroentangled fabrics, now, in many ways, needle felts would appear to be the ideal alternative for filtration, combining the possibility of greater flexibility and versatility in construction.

The performance of needle felted fabrics as filter medium has been strongly influenced by their structural features [62,154]. Filtration performance can also be affected by the fiber properties through their geometrical properties, surface finish, and electrical and mechanical properties. Synthetic fibers, particularly polyester fibers, are predominantly used for filter bags. Proper selection and designing of filters involves the understanding of the following parameters:

- *Physical characteristics of the dust*: Particle size distribution, particle shape.
- *Chemical composition of the dust*: Flammability, alkaline, acidic, corrosive.
- *Chemical composition of the gas*: Oxygen content, moisture, corrosive.
- *Operating temperature*: Ambient, high temperature, fluctuating temperature.
- *Mode of operation*: Process or nuisance collector, cleaning philosophy, controls.

Conventional needle-punched nonwoven fabrics (Figure 27) had a limitation in filtration as these failed to satisfy the recent strict environmental standards. The fabric is generally acceptable for collecting large particles (>5 μm); however, it is not equally effective for smaller, submicron-sized particles. The peg holes created in the needling process might allow the particle penetration [129]. On the other hand, fine particles with a diameter smaller than 2.5 μm ($PM_{2.5}$) are known to have the highest impact on human health because they can penetrate deeply into the human respiratory system. As environmental pollution is accelerating, standards about air-discharge pollution have become stricter. Problems associated with conventional filters include higher emissions, higher pressure differential (ΔP) across the filter due to clogging of the felt, and 'puffing' just after the cleaning cycle. Puffing occurs when particles are dislodged from the felt during cleaning, resulting in a temporary increase in emissions. For the improvement of filtration of filter bags, workers

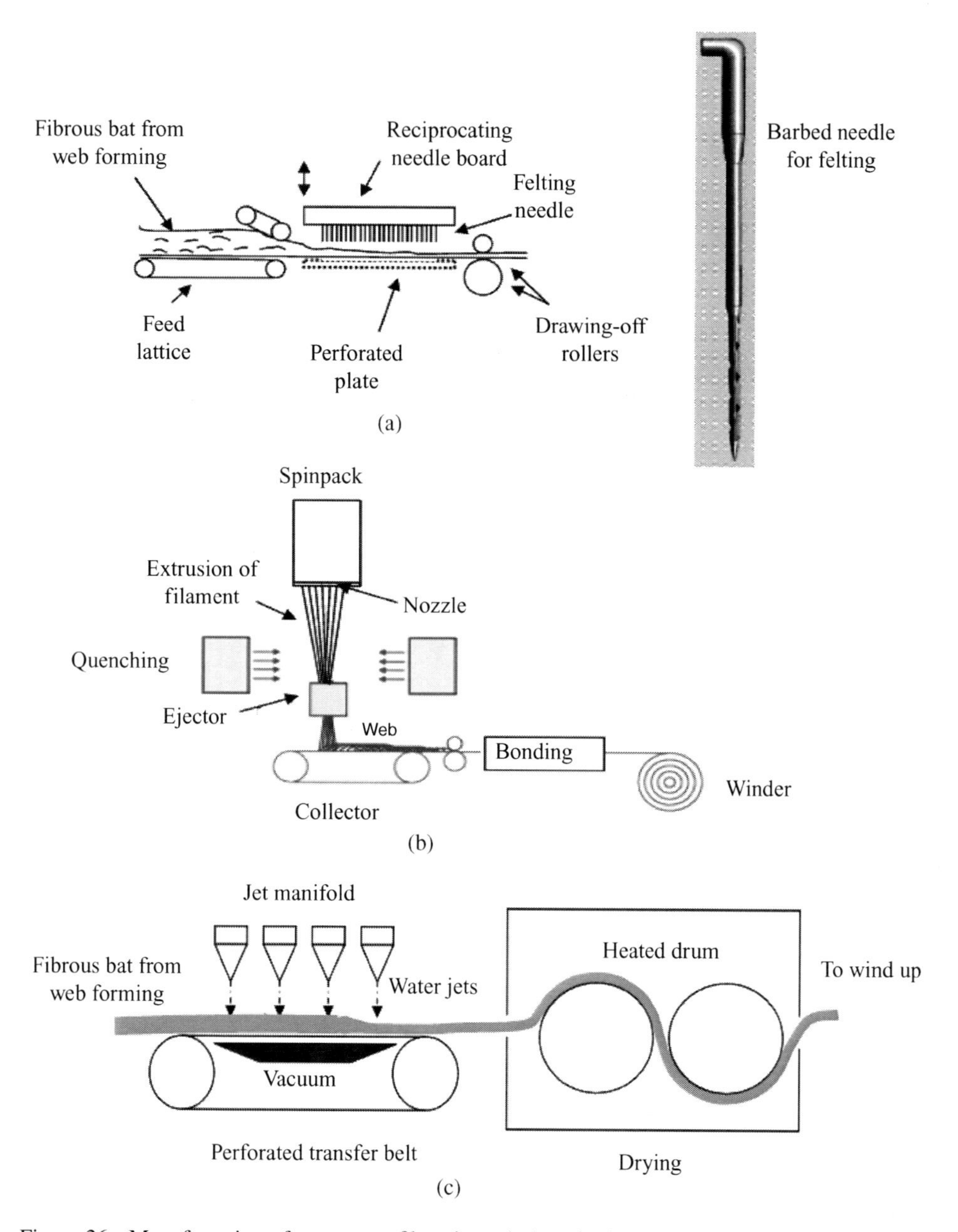

Figure 26. Manufacturing of nonwoven filter through three basic processes. (a) Needle felting; (b) Spun-bonding; (c) Spun-lacing (hydroentangling).

and researchers have investigated the various methods, such as addition of special fiber and membrane layers, and the finishing with coating or calendering [108,155–162]. Many of the techniques are in practice.

Adequate surface finishing may highly influence filtration, the type of filter cake, cleaning behavior, and finally the lifetime of the filter. Finishing may be based on thermal (heat setting, singeing, calendering, and condensing), chemical (antiadhesive, hydrophobe/oleophobe, antistatic, antiflammable, chemical resistant, etc.), and physical processes as well as the combination of the above [163]. Use of coated finish and application

Figure 27. Cross-section of a needle felt nonwoven showing the scrim and also the fiber re-orientation caused by the needling [129].

of membranes, in particular, have become common in the manufacturing of filter fabric [13]. Each of the processes is meant for a specific purpose depending on aerosol characteristics. For example, a heat procedure stabilizes any nonwoven fabric to a more stable product. By means of calendering, singing, or chemical process, a fabric surface can be modified, which helps in dust dislodgement during cleaning [90]. The fabric is usually given a cylindrical shape with appropriate seam construction. Development of seamless bags is also reported, which can be made through welding or by supporting the base material through screen. Fabric without seam can provide better filtration performance. This construction ensures that nothing bypasses the process media through holes in the fabric created from sewing the material. While designing the filter elements, a thorough understanding of environmental regulations, nature of particulates, and state of aerosol conditions is needed. Four factors related to the design of filter elements are:

- material of the fabric,
- fabric type and specification,
- fabric finish and structural modification, and
- filter element dimension and fabrication.

In many industrial applications, gas emitted by the process has a high temperature that may not be conducive to be treated directly in a collector. Since high-temperature filtration technology is rather expensive due to high costs of construction materials and filter elements, high-temperature filters will be economical if substantial advantages in process technology are expected. Typical reasons for the installation of a hot gas filter are, for instance, avoidance of condensation or subsequent wet processing, or protection of the installations like heat exchangers and wet scrubbers. An interesting option may be the avoidance of de novo synthesis of dioxins in the temperature range of 200 to 500°C by the removal of catalytic-active particulates above 500–600°C. Industrial practice shows that high-temperature filtration is seldom economical if heat recovery is the only benefit. In these cases, it is often cheaper to cool the gas below 250°C and to use the standard filter technology. In some specific cases, the gas temperature is lowered with water or through dry cooling in an air-to-air heat exchanger. In some applications, ambient air, often called dilution air, is introduced into gas to reduce its temperature [52]. High-temperature filtration is an option when requirements of the process demand the removal of particulate matter at elevated temperatures [71,117]. In contrast to the above, in certain instances, gas is to be heated to avoid condensation. This can be accomplished through the introduction of hot air into the system.

In the baghouse design procedures, there is a large variation in the recommended appropriate values for air-to-cloth ratio, elutriation velocity, bag spacing, and dimensions. Other key design parameters are the choice of filter medium, operating differential pressure, and baghouse footprint. A systematic study could entail a proper designing of the system. In a recent work, the design of pulse-jet baghouses for fine collection has been investigated [164]. Firstly, the key design parameters and recommended values for these baghouses are presented, showing a large variation among sources. Previously reported design guidelines are compared for milk powder baghouses and their suitability assessed in the light of practical experience. A procedure for baghouse design has been developed that differs from previous work as it includes the area outside the bag bundle as a key design parameter and uses an optimization routine to solve the system of equations. Significant variation has been observed among the four design procedures found to determine the air-to-cloth ratio when applied to baghouses collecting milk powder. The air-to-cloth estimate by the method given by Lõffler et al. [165] is more specific for fine collection in milk powder plants, and is highly recommended.

5.2. *Material of the fabric*

5.2.1. Material type

The selection of proper fabric is one of the primary factors for the proper functioning of the fabric filters. A variety of common fibers like natural cotton, wool, etc. or man-made fabrics like polyester, acrylic, polypropylene, polyamides, glass, aramide, polytetrafluoroethylene (PTFE), etc., are being used for the manufacturing of filter bags [13]. In the market, many manufacturers offer various types of fiber materials, either named after used chemicals or in their own trade name. Properties of various common fibers and their uses in the industry are given in Table 9. Selection of a particular fiber material in a specific application depends on various properties of fiber like heat resistance, chemical resistance, resistance to moisture, physical properties, relative price, etc. If the fiber material is not properly selected, there could be many consequences, such as severe shrinkage of material, fiber degradation following embrittlement of structure, accelerated pulse-flex fatigue, etc. However, the worst affected bags do show severe flex fatigue at the top of the bag with apparent but less significant flex fatigue at the bottom of the bag.

High-temperature filtration is one of the most promising developments in particle-collection technology. Process gases generated at high temperatures and/or pressures might contain small particles along with various harmful gaseous components. If these gas streams can be cleaned at elevated temperatures and pressures, then processes can be made more efficient in terms of energy and more integrated in terms of process technology. The long-term needs of power-generation systems have driven this development, but the focus has now shifted more to the chemical and process industry. If the temperature of the gas stream is higher than the one sustained by the fiber, and cost considerations preclude the possibility of gas cooling prior to dust collection, then the alternative means of collection will have to be sought. Depending on the duration of exposure, high temperatures may have several effects on the fiber, the most obvious is the loss in tenacity due to oxidation and less effective cleaning due to cloth shrinkage. The maximum operating temperature for each fiber is lower than the melting point [90].

Table 9. Properties of fiber material used in filter bags.

	Maximum temperature (°C)			Physical resistance					Chemical resistance							
Fiber material	Continuous	Intermittent	Under moist condition	Dry heat	Moist heat	Abrasion	Shaking	Flexing	Hydrolysis	Mineral acids	Organic acids	Alkalis	Oxidizing agents	Organic solvent	Relative price	Application examples
Cotton	80	—	80	G	G	F	G	G	—	P	G	F	F	E	—	Cement industry, pharmaceuticals, food industry, fiber board plant, wood plant (presently use is limited)
Wool	90	120	105	F	F	G	F	G	—	F	F	P	P	F	—	Cement industry, food industry, fiber board plant, wood plant (presently use is limited)
Polyester	135	—	100	G	F	G	E	E	P	G	G	F/G	G	E	1.0	Mining, cement, iron and steel, wood, ceramic, plastic and pigment
Nylon (polyamide)	110	—	100	G	G	E	E	E	—	P	F	G	F	E	—	Pharmaceuticals, food industry, fiber board plant, wood plant (presently use is limited)
Polyacrylic	125	—	120	F	F	F	P/F	G	G	G	G	G	G	G	1.6	Asphalt, spray dryer, lime, plastic
Polypropylene	90	120	90	G	F	E	E	G	E	E	E	E	G	G	1.0	Food industry (milk powder, sugar flour), detergent
Nomex	190	230	170	E	E	E	E	E	F	P/F	E	G	G	E	5.0	Asphalt, iron, cement/lime kilns, metal alloy smelting, ceramic
Teflon (PTFE)	260	290	260	E	E	P/F	G	G	E	E	E	E	E	E	15.0	Carbon black, coal-fired boilers, incineration, cupola, ferro/silica alloy furnace

(Continued)

Table 9. Continued. Properties of fiber material used in filter bags.

Fiber material	Maximum temperature (°C)			Physical resistance					Chemical resistance							
	Continuous	Intermittent	Under moist condition	Dry heat	Moist heat	Abrasion	Shaking	Flexing	Hydrolysis	Mineral acids	Organic acids	Alkalis	Oxidizing agents	Organic solvent	Relative price	Application examples
Glass	260	315	260	E	E	P/F	P/F	F	G	E	E	G	E	E	2.5	Coal-fired boilers (FBC, PC), electrosmelting oven, cement/lime kilns, industrial and small municipal boiler applications, furnace for metal melting, incinerators
Ryton (polyphenylene sulphide)	190	—	150	E	E	G	G	G	E	G	G	G	P	E	5.0	Coal-fired boilers (FBC, PC), electrosmelting oven, furnace for metal melting, biomass, waste incinerator
Polyimide P84	260	—	—	E	—	G	G	G	F	G	G	F	—	—	6.0	Cement kiln, metallurgy, biomass, waste incinerator, coal-fired boilers
Ceramic Nextel 312	1150	—	1150	E	E	—	—	—	E	E	E	G	E	E	—	Pressurized fluidized bed combustion (PFBC), integrated gasification combined cycle (IGCC), and fuel cell technologies

E = Excellent, G = Good, F = Fair, P = Poor.

In certain applications, the presence of moisture in the hot gas stream (at above 100°C) will convert into a superheated steam, which in turn will further aggravate the situation. This will also cause rapid degradation of many fibers through hydrolysis, the rate of which is dependent on the actual gas temperature and its moisture content. Furthermore, fiber material can be seriously affected by chemical nature of the gaseous compound; the problem can be further compounded in the presence of heat and moisture. For example, during combustion of fossil fuels, the sulfur that is present in the fuel oxidizes in the combustion process to form SO_2, and in some cases, SO_3 is liberated. In the presence of moisture, SO_3 will be changed into sulphuric acid and if the temperature in the collector is allowed to fall below the acid dew point (in excess of 150°C), rapid degradation of the fiber will ensue. Polyaramid fibers are particularly sensitive to acid hydrolysis and, in situations where such an attack may occur, more hydrolysis-resistant fibers, such as those produced from polyphenylenesulphide (PPS), would be preferred. Therefore, it is essential to choose the right kind of fiber materials depending on applications (Table 9).

Normally, a high proportion of fabric dust collectors do not face such thermal or chemical constraints. The commonly used fiber in dust collection is of polyester origin, this being capable of continuous operation at a reasonably high temperature (150°C), and is priced competitively. However, if there were a chance of hydrolysis, fibers from the acrylic group would be the preferred choice [90]. On the other hand, for the simultaneous removal of particles and gaseous compounds in room environment by fibrous filters, Otani et al. [166] proposed fibrous filters made from carbon fibers with a chemically activated surface. It was shown experimentally that the penetration of charged particles decreases by almost an order of magnitude for activated fibers compared to untreated ones. Otani et al. [166] have also tested the process of conversion of organic gases to particles as a means for removing them by fibrous filters. It may be added that for low-resistivity dust, there is no difference between nonconductive bags and bags with conductive yarns. For dust of high resistivity, the cleaning effect is better when the bag is conductive [167].

According to Dickenson [168], the development of heat and corrosion-resistant fabric medium, particularly modern treated needle felts, has been one of the most significant advances in the dust extraction and treatment of hot gases. However, during the choice of fiber material at high-temperature application, it is important to see whether the effect of temperature is continuous or intermittent. It is important to note that the application of conventional high-temperature filter materials in the temperature range of 250°C and higher is substantially restricted. For example, common fiber materials used in pulse-jet filter are homopolymer acrylic (Dralon-T) at temperatures less than 140°C, polyphenylene sulphide (Ryton), aramid (Nomex), etc. at a temperatures of about 200°C, and Polyimide (P84) and Teflon at temperatures of about 250°C. It may be added that, in spite of polyimide and fluorocarbon fibers having their melting and decomposition temperatures in the range of 400–650°C, these show significant shrinkage even in the much lower temperature range (280–300°C). This puts a limit on the maximum workable temperature.

In a report published in 1992 [75], analysis of various fabrics used for filtration in coal-fired boilers provides a fairly good idea about the behavior of fabrics under similar conditions (Table 10). While designing a bag filter, cost is also likely to be an important criterion. Report states that properly designed and built filter with appropriately selected fabric showed good average life of three years or more. Following is a comprehensive view of the performance of different high-performance fibers [52,75,100,169–172]:

- Usually glass fiber materials can be operated at 30–260°C, but they have certain shortcomings. Huyglas type is specially engineered felt designed to withstand temperature

Table 10. Operational observation of various fiber material used for filtration in coal-fired boilers [75].

Fabric composition	Operational observations
Polyphenylene sulphide – PPS (Ryton) The needle felt made out of PPS is backed with PTFE/PPS scrim	• Ryton fiber is exceptionally resistant to acidic conditions; however, not chemically inert and may be degraded by some flue gas conditions. • Ryton showed a very good life ranging from three to four years. One particular installation showed even six to seven years of life
Aramid (Nomex)	• According to DuPont guidelines, Nomex must not be used if any two of the stated conditions are exceeded during fabric filter operation: (i) 160°C (325°F), (ii) 15% moisture by volume, (iii) 350 ppm SO_2. • Nomex fiber has poor resistance to acid hydrolysis. It should not be used in pulverized coal-fired, and stoker-fired boilers burning coal having more than 0.3% sulfur. • Nomex bags should not be exposed to untreated or improperly treated flue gas in spray dryer or FBC baghouses. • Nomex exhibited average life of three to four years downstream of FGD or FBC.
Homopolymer acrylic (Dralon T)	• Appears to be inexpensive and effective alternative. • Continuous temperature must be kept below 125°C (260°F). • In Munmorah power station, New South Wales, the filter bags exceeded three years guaranteed period with only three bags failure.
Polytetrafluoroethylene – PTFE (Teflon)	• Highly durable due to inherent chemical inertness. • The length of uninterrupted service is dependent on the tendency to blind resulting in high pressure drop or emission, or both. • To extend the bag life, DuPont has come out with a fabric called Tefaire, which is a blend of Teflon and fine glass fiber.

excursions up to 316°C and has a continuous operating temperature of 287°C. Needle-punched glass fiber filter materials (Huyglas) need special fitting and arrangement on the frames to ensure long-life performance. Membrane filters made from glass fabrics covered with porous fluoropolymer have low permeability, which results in higher operating costs for filtration. Moreover, the most abundant E-glass fibers have poor acid and alkali resistance.

- Aramid (Nomex) fibers possess good chemical resistance compared to other fibers, especially at 150–180°C (300–350°F). It also has flex resistance properties for utility baghouse services but has proven vulnerable to degradation in some flue gases, depending apparently on SO_x concentration. As mentioned, aramid felt get hydrolyzed at above 150°C due to the presence of moisture and acids in the gas. Hydrolysis is a reaction where water reacts with the substance to produce a new substance. Presence of acids in the gas enhances the reaction and results in the damage of felts. However, addition of dry flue gas desulphurization upstream of the baghouse makes the operating environment much more compatible with Nomex fabric, which can also be given acid-resistance finish.

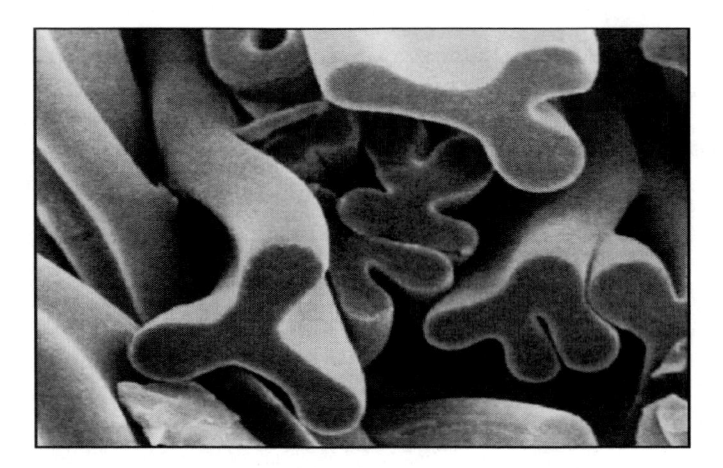

Figure 28. Multi-lobal structure of P84 fiber [172].

- PPS generally has excellent chemical resistance relative to other fibers. PPS has advantages regarding stability against condensing acids, which is of interest in discontinuous operated filters where condensation takes place, or in a situation like a coal-fired boiler, which operates close to the acid dew point. However, PPS can degrade under a variety of complex reactions that, to date, are not completely understood. Reactions can occur between polymer chains and oxygen, and may react with sulfur atoms to form sulphones and sulphoxides. The combination of heat and aggressors, such as HNO_3, HCl, and H_2SO_4, further accelerates and complicates this degradation mechanism. In situations where there is continuous exposure to temperatures greater than 190°C and atmospheres having more than 15% oxygen, a more costly material such as polytetrafluoroethylene (PTFE) is preferred.
- P84 is a more heat- and oxidation-resistant material in comparison to PPS (i.e. a higher bag life is expected if oxidation is the limiting factor). Multi-lobal structure of P84 fiber (Figure 28) results in a more porous and thicker filter medium than the one built from circular fibers. Its unique structure also offers high-filtration efficiency. Furthermore, the needle felts of P84 fibers are more resistant against mechanical stress at higher dust loads of above 500 g/Nm^3 or abrasive particles. Due to the good chemical resistance, the use of alternative fuels and the variation of flue gas conditions will not cause problems for P84. If it is not suitable to use 100% P84 technically or economically, then its combination with other materials (e.g. PET, PAN, PPS, PTFE) can utilize the high-filtration efficiency. Various structural combinations such as P84 needle felt with P84 scrim, P84 on a PTFE scrim (not PTFE membrane), or on a hybrid scrim where P84 and PTFE yarns are used for scrim construction are reported.
- Fabrics made from polytetrafluoroethylene (PTFE) fibers demonstrate excellent performance for hot corrosive gas filtration, but their use is limited due to their very high costs. In addition, PTFE fibers also shrink at elevated temperatures.

In an attempt to develop material for high-temperature applications, basalt-based composite filter fabric has been found to be quite successful [173]. The materials combine the filtering characteristics of needle-punched fabrics made from high-temperature synthetic fibers (P84) with high chemical resistance, thermal resistance, and strength of basalt fabric as scrim. For composite fabrics, the shrinkage of the nonwoven layer does not lead to the deterioration of filter ability or changes in the dimensions of filter bags, and the bags can be operated at temperatures of up to 300°C. The service life of both P84/basalt filter composite (Meteor bags) and membrane-laminated glass fiber bags is comparable. Basalt composite filter fabrics can also be successfully used for cleaning corrosive hot gases, or waste air containing hot particles having temperatures of over 800°C. Meteor fabric was reported to be used for cleaning air with 1500 ppm of SO_2 at 160°C and high-moisture content in South African Republic; filtration of waste gas after nickel film production from carbonyl of nickel in Canada; as well as for filtering stack gas of molybdenum powder production in Russia. There is also a report of significant increase of lifespan of basalt composite filter bags through *LD-Mobile/HEC* monitoring and cleaning technology [174].

The decision regarding the most suitable material needs a more detailed view on the particular application than just finding a fabric, which is chemically and mechanically stable. For illustration, choosing suitable material in cement kiln plants needs critical analysis of the process. Depending on the availability of cooling systems, the baghouse inlet temperature in most cement kiln plants is between 100°C and 200°C. Furthermore, in addition to acidic flue gas components, such as SO_x, HCl, and HF, the high content of NO_x due to high combustion temperatures, is typical for kiln off-gases. Under such an environment, a wide range of filter fabric materials is used. Polyacrylic homopolymer (PAN, e.g. Dolanit) is suitable when peaks above 140°C can be prevented; Polyphenylene sulfide (PPS), m-aramide, and polyamide-imide cover peak temperatures up to 200°C. Continuous operation at a temperature of 180°C or above, or peak temperature above 250°C, lead to an insufficient lifetime for these materials. P84, PTFE, and glass are most capable of withstanding such operating conditions. During kiln dedusting, one of the most important ways is how the acids' dew point crossing is handled and how the baghouse is preserved during standstills. In case the operating temperature goes below the dew point, it is important to stop pulsing to keep the protective layer cake on the bag surface as much as possible, as the cake on the surface will neutralize acidic condensation. Secondly, it is important to keep excursions at the maximum temperature limit as short as possible. The nature of materials used for the filter fabrics is such that they can only survive a limited period under the maximum operating temperature. Shutdown, bypassing, or cooling water injection prior to the baghouse are the measures that can obviate the risk [172].

When kiln and raw mill dedusting is combined, the temperature within the operating mill for most of the time (in 90% cases) is typically between 90°C and 160°C. The choice of suitable fabric materials is restricted because when the mill has stopped, the filter fabric must withstand temperatures up to and above 200°C. However, glass fiber fabric with PTFE membrane may not be able to fulfill adequate performance standards. The high dust content when the mill is in operation often leads to abrasion and membrane failure. The gas composition in a compound operation differs mainly by the higher water content from kiln off-gases. P84 needle felts are widely unaffected by abrasion and are capable of staying below the strictest emission limits. In such applications, a lifetime of three to six years is commonly achieved with P84 felts (preferably with layered finer fiber component at the upstream side), granting process stability, low emissions, and low differential pressure during the entire bag life. Another advantage of P84 is that spare bags stored over an extended period of time for immediate replacement in case of failure are not affected in quality. It may be added that baghouses are typically designed with air-to-cloth ratios of 0.6–1 m/min in comparison to values slightly above 1 m/min in case of more advanced felts. As the public concern increasingly covers other pollutants like SO_x and NO_x, additional flue gas treatment technologies will find their way to cement applications. Technologies containing $DeNO_x$ processes, as well as the absorption of SO_x, are available for other applications but

Table 11. High-temperature process in the industry [71].

Industry	Type of process	Temperature
Power generation	Pressurized coal combustion	Up to 1500°C/25 bar
	Gasification	Up to 1000°C/50 bar
Handling of residues (Incineration/gasification)	Municipal industrial waste, biomass, plastics	Up to 1000°C/1 bar
	Soil decontamination	Up to 1000°C/1 bar
Chemical process technology pp	Catalytic crack processes	300–500°C/<5 bar
	Calcination/drying	Up to 1000°C/1 bar
	Particle production in the gas phase (e.g. TiO_2, SiO_2, carbon black)	Up to 1200°C/<5 bar
	Fluidized bed process	Up to 800°/1 bar
Metal process industry	Metal roasting	Up to 1000°C/1 bar
	Smelting process	Up to 1000°/1 bar
Ceramic industry	Ceramic production	Up to 800°C/1 bar
Glass industry	Glass production	Up to 800°/1 bar

so far are not widely used in the cement industry. Again, a fabric filter unit will play a key role in these processes because it can act as adsorbent and absorbent of gaseous pollutants.

Although high process temperatures much above 300°C, are beneficial for cycle efficiencies, they impose severe limitations on the type of filter medum to be used and on the mechanical durability and corrosion resistance of components in the gas-cleaning unit. Table 11 highlights the plausible applications of filters at very high temperatures in the industrial process [71]. In general, for high-temperature gas-filtration systems, several types of filter elements, such as filter fabric made out of mineral fibers/metal fibers/incotel (austenitic nickel-chromium-based super alloys)/glass fiber, silicon carbide candles, and ceramic composite and ceramic fabric filters, can be used. The most efficient way of separating particles at high temperatures (up to 1000°C) is to use ceramic filters (rigid/flexible). They are unrivalled in the filtration of corrosive gases with temperatures of more than 500°C, but when filtering gases with temperatures less than 300°C, they possess relatively lower strength and air permeability, even compared to fabrics based on glass fibers [100,114]. A range of metallic filters can be formed into flexible fabrics analogous to polymeric needle felts – and rigid screen-type filters based on metallic meshes [100]. Metal fibers can be used for a temperature range of 300–600°C. This type of filter medium is, however, relatively expensive and does not have temperature rating as high as that of the ceramic fibers, hence, is not used widely in pollution control applications. However, in all these cases, bag construction methods are important, seam failure being a significant source of bag failure. Seamless construction of ceramic bags eliminates sources of failure (Figure 29) [81]. The nonwoven media can be sandwiched between the supporting grids knitted or woven from either stainless steel or ceramic fibers [71,77].

In an experimental investigation [71], a ceramic and two types of metal high-temperature filter elements (sintered fleece and woven bags) were tested on a laboratory pilot filter with different filtration velocities up to 2.5 m/min and temperatures ranging between ambient and 600°C. Operational differences between ceramic and metal filter media were derived from the measurement of the pressure drop behavior as well as the residual pressure drop of filter elements. In the case of metal filter elements, an additional parameter, i.e. the change of media structure dependent on temperature, is an important parameter to be considered

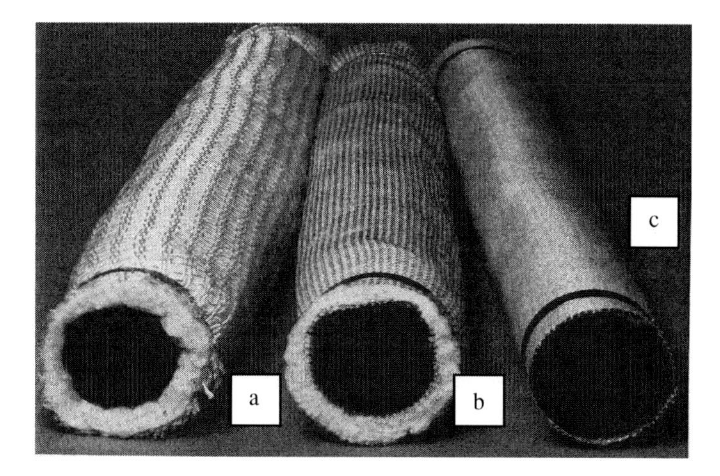

Figure 29. Seamless ceramic filter bag. (a) Nonwoven ceramic media with metal screens; (b) Nonwoven ceramic media with ceramic fabric screens; (c) Woven fabric ceramic filter media [81].

[71]. However, for long-term stable operation at high temperature (600°C), a well-designed candle filter working under pulse-jet-regeneration system is very common.

5.2.2. *Fiber fineness and cross-sectional shape*

Fiber fineness in a filter element varies widely. The fibers are normally in the range of 1.66–3.33 decitex, though trends of considerably finer 'microfibers' (e.g. less than 1 decitex) have gained some prominence. Using mirodenier fiber can bring down the pore size from 35–66 to 12–25 μm. Higher filtration efficiency can be obtained due to larger surface area and smaller size of pores. However, processing finer fibers requires proper attention during the formation of fibrous batt (require higher carding) and at the time of fiber consolidation through needle-punching operation. Production of microfiber can be achieved by means of the so-called splittable fibers. Such fibers comprise a number of elements, which are bound together at the extrusion stage. However, because of the subsequent mechanical action of carding or aqueous treatment in the case of water-soluble binders, the individual elements split from the parent structure to produce a microfiber web [13,90,163]. In a study, use of coarser fiber (7 denier) instead of 2.7 denier/3 denier fiber in PPS felt has been found to improve the performance of a hybrid filter unit (**CO**mpact **H**ybrid **PA**rticulate **C**ollector – COHPAC) as permeability increases substantially without compromising particulate collection efficiency. Continuous cleaning resulting from heavy ash accumulation has not been a problem unless the performance of the upstream hot-side electrostatic precipitator degrades [175]. However, the general trend is to use finer fiber for achieving higher filtration efficiency. Some manufacturers adopted layer fabric concepts with high particle-collection efficiency fibers on the surface whilst retaining coarser, less expensive fibers on the back for maintaining lower pressure drop without compromising the filtration efficiency.

Although most of the fibers utilized in dust collection are of circular cross-section, irregular, multi-lobal-shape fibers, (e.g. Lenzing's/Evonik's P84) and peanut shape fibers (e.g. DuPont's Nomex) are also possible. The latter possess a higher surface area with potentially superior particle-collection capability [90,172]. In an early study [176], it was observed that at any level of fabric weight the hollow fiber fabric shows higher air permeability, followed by trilobal and round fiber fabrics. However, the filtration efficiency is less for hollow fiber fabrics followed by round and trilobal fiber fabrics of similar fabric weight.

5.3. Fabric types and specifications

Fabric types and specifications mainly depend on equipment requirement, application, and filtration-specific requirement. In most of the cases, fabrics with distinct surface characteristics are used in both scrim-supported and self-supported styles, and with both sewn seams and fused seams. Fabric specifications could vary depending on the intensity of pulse-jet pressure, its frequency, and duration in the equipment. Various requirements of fabric filter can be enlisted as follows:

- Filtration requirement, which encompass smallest particle retained, overall filtration efficiency, resistance to flow, tendency to blind, and cake discharge characteristics.
- Physical properties, such as dimensional stability, elongation at break, bursting strength, resistance to creep/stretch, resistance to wear, absorption of moisture, electrostatic charge, flexing strength.
- Heat resistance.
- Chemical resistance, such as acid, alkali, reducing agents, organic solvent, etc.
- Ability to be fabricated, sealing, and gasketing function are also important.

A needle-punched fabric with weight of 450–500 g/m^2 is very commonly used, which provides adequate strength and life to filter element. However, fabric with weight of 550–600 g/m^2 with equivalent pore diameter of 40–50 μm for abrasive and fibrous dust and 25–30 μm for exceptionally high-collection efficiency are also used [13]. Material of higher weight usually leads to greater thickness. A thick material with small equivalent pore size gives better filtration efficiency. On the other hand, material with higher thickness decreases the flex, making the cleaning difficult. Higher fabric thickness also leads to higher pressure drop across the fabric [177]. Increase in fabric weight decreases air permeability and therefore pressure drop can rise largely during filtration [176,178]. Therefore, fabric weight should be judiciously selected depending on the filtration requirement.

In designing the nonwoven fabric, material consolidation and hence the pore size and porosity can be regulated through needling parameters – (i) needle design, (ii) needle fineness, (iii) needle orientation, (iv) needle board pattern, and (v) needling program, that is, punch rate and penetration. With the increase in material consolidation, filtration efficiency can also be increased but at the cost of higher pressure drop. To improve the integrity of the material, physical requirements during cleaning of most needled felts, notably in Europe, scrim is used. In the United States, approximately 50–75% of the filter materials from needle felts are used without scrim [179]. Inclusion of scrim also improves filtration efficiency but at the cost of higher pressure drop [176,180]. Depending on the tensile specification of the finished needle felt, the area density of scrims is usually in the range of 50–150 gm^{-2}. The design of the scrim is frequently 'over-engineered' to compensate damage during the needling process, and may be alleviated by the judicious selection of needling parameters.

The needles themselves, typically 75–90 mm in length, are mounted in a board, the arrangement or pattern of which is so designed as to provide a surface which is as uniform as possible and devoid of 'needle tracking lines'. Normally triangular in cross-section, the needles contain a series of barbs, which are set into the corners. Typically, a needle contains nine barbs, which are designed to engage the fibers on the downward stroke of the punching action, yet emerge completely clean on the upstroke. Hence, the fibers become mechanically locked both to other fibers in the fibrous batt and to the woven scrim. The barbs may be regularly spaced over the length of the needle blade, or more closely spaced for more intensive needling and the production of a more dense structure. In another design, the barbs are located on only two of the three corners, this style being used where maximum protection to one of the scrim components is required. The density of the needles in the needle board, the frequency of needle punching, the style of the needle, and depth of penetration through the structure have significant influence in controlling the thickness and density of the final assembly, and also the strength retained by the scrim. However, with improvement in needling technology, uses of scrim have become limited [90]. It is observed that air permeability decreases with the increase in needling density and needle penetration up to a certain limit, but at higher levels of needling density (400 punches/cm^2) and needle penetration (14 mm), the air permeability of the fabric increases [178]. This is probably due to fiber rupture/damage caused by higher needling action resulting in holes in the structure.

Glass fibers are of brittle nature and therefore not usually amenable for the production of nonwoven fabric. They are mostly used in woven filter materials, which can be used in pulse cleaning with compartmentalized design with a gentle cleaning technique. Although a heavier version (750 g/m^2) of glass fiber woven fabric is capable of achieving lower emission, it exhibits higher pressure drop under pulse-jet filtration situation. Therefore, pulse cleaning with woven fiberglass has to be sized more generously with lower filtering velocity than what is used in felt fabric. For the replacement of glass fiber for high-temperature application, different aramide fibers can be used in nonwoven filter materials for pulse-jet filtration unit.

In addition to the needle-punching process, spun-bonded, chemical-bonded, and hydroentangled fabrics are also used. In case of thermal bonding and latex bonding, the binder can provide shape stability at the pleats. Thermal bonding by using bi-component fibers is constantly increasing [163]. The combination of needle-punching and thermal bonding processes is useful for reduction in the differential pressure of the bag filter media and improving the collection efficiency. In a study [181], optimum level of performance was obtained at the ratio of 30% in a fiber mixture, the main-needling stroke of 1200/min, the process speed of 2 m/min in a needle-punching machine, and the thermal bonding temperature of 180°C. It revealed that many needling strokes resulted in longer lifetime of filter. Dura-Life™* polyester filter bag media [182] is engineered with a unique hydroentanglement process, which is claimed to create a filter bag that has smaller pores, better surface loading, and better cleaning resulting in increased bag life. With the development of new types of fabrics, specification and functional requirements have changed to a large extent.

5.4. *Fabric finish*

Through the application of finish, basic fabric characteristics can be improved for efficient filtration. These are designed essentially to improve the following characteristics [13,81,127,156]:

- Fabric stability;
- Surface characteristics;
- Permeability of the fabric;
- Filtration collection efficiency;
- Cake release characteristics;
- Resistance to damage from moisture and chemical agents;
- To restrict the dust particles to the surface of the fabric so as to reduce the blinding tendency.

A number of finishing processes are employed to achieve these goals, e.g. heat setting, singeing, raising, calendering, special surface treatments (membrane lamination and coating), and chemical treatments (hydrophobic finishing, flame retardant finish, and antistatic treatment). Most of the finishing processes are very common for many textile applications except the special surface treatment through incorporation of a more efficient membrane in a lamination operation or by coating of filter media. This is a remarkable development in filtration science through which high levels of filtration performance have become plausible [81,156]. Apart from enhancing filtration efficiency, membrane and coated fabric offer the following advantages:

- Reduced emission of finer particles;
- Better cake release property through imbibing smooth surface and moisture-repellent property of the fabric;
- Enhances the life of filter due to reduced pore blocking by the dust particles.

5.4.1. General

The most common types of finish applied to filter fabrics are heat setting, singeing, and calendering. Fabrics are also chemically treated depending on the specific requirement. The basic purpose of heat setting is to improve stability from thermal shrinkage. If the thermal environment during filtration is above the glass transition temperature of the constituent fiber, onset fabric will tend to shrink. In a pulse-jet fabric collector, lateral shrinkage could result in the fabric becoming too tight on the supporting cage, leading to inefficient cleaning and ultimately an unacceptable pressure drop. Furthermore, due to shrinkage at the time of filtration, reduction in pore size will occur, which might anchor trapped particles more rigidly and makes the cleaning difficult. In the heat-setting process, the fabric is treated either in a steam atmosphere or in a dry heat environment. In order to ensure stability during use, the temperature in the heat-setting operation will invariably be significantly higher than the maximum continuous operating temperature of the material in question. Furthermore, since complete fiber relaxation is a temperature–time related phenomenon, manufacturers will also process at speeds that are appropriate to achieve the desired effect. In addition to stabilizing the fabric, the heat-setting process will also increase the density of the structure through increased fiber consolidation. This in turn will further assist in achieving a higher level of filtration efficiency [90].

Singeing is another important finishing process intended to remove protruding fibers from needle felt made out of staple fibers. The projected fiber may inhibit cake release by clinging to the dust. This is achieved by passing the fabric at relatively high speed, over a naked gas flame or, in another technique, over a heated copper plate. Singeing conditions (i.e. speed and gas pressure) will normally be adjusted according to polymer type and the intensity required by either the end-use application or the individual manufacturer's preference [90].

Calendering is a mechanical process that owes its effectiveness to the application of factors such as heat, pressure, and moisture to the fabric to plasticize the fibers through consolidating its structure and thereby enhancing its physical properties [183]. The calendering operation not only regulates density and permeability of the fabric but also improves its surface smoothness. The treated fabric will therefore exhibit higher filtration efficiency and dust-release properties.

Membrane filters are widely used to sample radioactive aerosol or microorganisms. Experiments on a high-velocity pulse-jet filter used to control emissions from a wool carbonizing process had shown that it is necessary to use filter cloths with a surface skin to prevent premature blinding of the fabric [82]. The skin partially prevents dust and wool fiber from penetrating into the depth of the fabric and, subsequently, the dust is more easily dislodged during cleaning. Although both Goretex® (PTFE membrane) and a calendered needle felt have given satisfactory performance at a filtration velocity of 65 mm/s, there is a gradual increase in the resistance to airflow through fabrics with time. This is due to the accumulation of dust on the fabrics, which is difficult to remove by the normal cleaning pulse. It was suggested that the efficiency of the cleaning process could be improved by cleaning the filter offline rather than the more normal online cleaning. Under these conditions, the dust that is usually re-deposited on the bag is allowed to reach the hopper [82].

In a study [184], surface-treated and untreated fabrics were investigated, including microporous coated, laminated e-PTFE, heated calendered, and surface-singed material. Surface treatment results in smaller mean pore sizes and higher cake resistance. Unlike other surface-treated material, singed fabric showed initial depth filtration followed by surface filtration. Although with surface filtration, markedly high-filtration efficiency can be obtained, it is at the cost of higher pressure drop.

Chemical treatments are normally applied for one of two reasons: (i) to assist in dust release, especially where moist sticky dusts, possibly containing oil or water vapor are encountered; (ii) to provide protection from chemically aggressive gases such as SO_2 and SO_3 emitted in the process. However, in the case of SO_3, it is possible that common chemical treatments will not be fully effective in the presence of moisture and hence a more chemically resistant fiber must be sought. Other chemical treatments may also be employed for more specific purposes. For example, proprietary treatments, usually involving silicone or PTFE, enhance yarn-to-yarn or fiber-to-fiber 'lubricity' during pulse or flex cleaning and similarly, where flammability is a potential hazard, padding through commercially available flame-retardant compounds may be necessary [90].

5.4.2. *Membrane filters*

In 1973, W L Gore & Associates pioneered the use of Gore-Tex® expanded polytetrafluoroethylene (ePTFE) membranes that are applied to a needle felt support for air/dust filtration applications. Figure 30 shows the photomicrograph of a PTFE membrane. Membrane provides higher efficiency than conventional fabrics. Its pores allow air to pass through while trapping very small particles below 2.5 μm. The efficiency of membrane filter is therefore higher than the conventional filter particularly for smaller particles (Figure 31). Due to the extreme thickness (12–75 μm) of the membrane, it is essential to combine it with a suitable substrate either by special adhesives or, where appropriate, by flame bonding. This membrane is attached to needle-punched felts, spun-bonded, or over spun-lace nonwoven. The substrate can be made out of different fiber polymers: homopolymer acrylic, torcon, nomex, static conducting polyester, polyester, or polypropylene. PTFE membrane can also be used over woven fabric of PTFE yarn. Since the PTFE membrane is chemically inert and can withstand high temperature (260°C), selection of the base material depends on operating temperature and chemical environment during filtration. Filters with PTFE membranes are engineered to unique microstructures to provide exceptional flow capability along with high efficiency [108,139,157,158,185,186]. The PTFE membrane laminated to the spun-bonded polyester media yielded slightly higher filtration efficiencies of 99.999% (99.9975% efficiency without lamination) while operating at a slightly higher pressure drop than the spun-bonded media. The spun-bonded media has been installed in baghouses

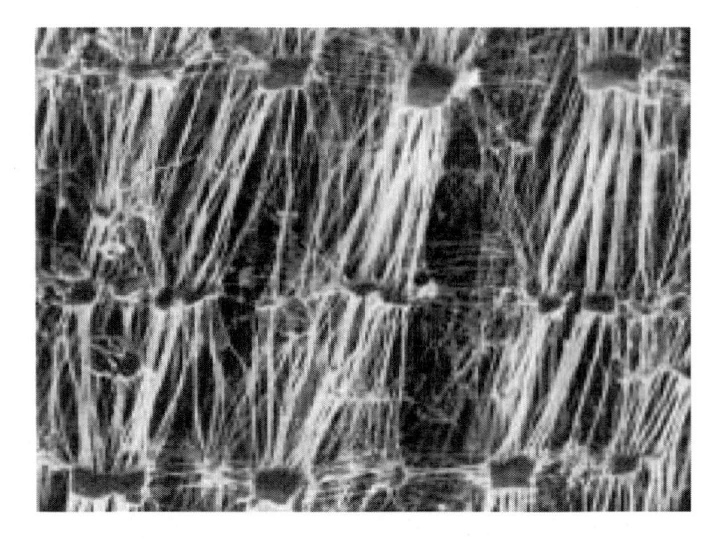

Figure 30. A microphotograph of expanded PTFE membrane.

ventilating many applications [185]. The numerous advantages of membrane filters, which are often highlighted by the product manufacturers, can be enlisted as follows:

- Reduce maintenance costs;
- Increase plant utilization;
- Provide longer filter life;
- Meet and exceed environmental compliance;
- Reduce particulate matter and down time;
- Reduce energy costs;
- Reduce cleaning cycles;
- Increase throughput.

Glass fiber fabrics without membranes cannot meet today's emission requirements. In a study [187], the performance of GORE-TEX® membrane/TEFLON® B fiberglass fabric filter bags on particulate emissions from a municipal solid waste combustion facility was

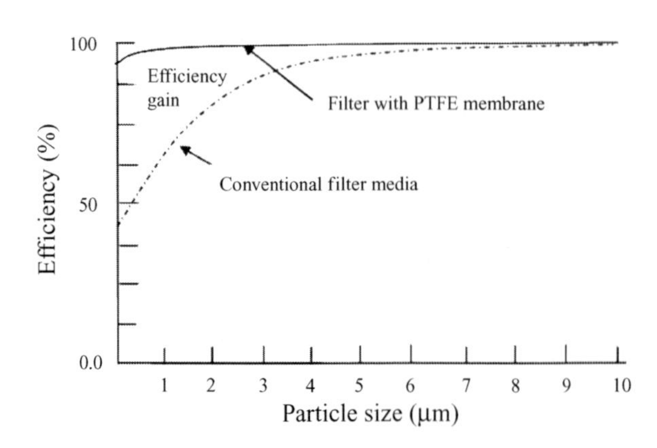

Figure 31. Comparison of fabric with and without PTFE membrane.

assessed. The technology consists of a pulse-jet fabric filter collector equipped with a lime slurry spray dryer absorber. Following are some of the attributable factors to the success of the facility's fabric filter collector: (1) superior operation of the boiler and spray drier absorber, (2) infrequent and low-pressure pulsing, (3) no penetration of particulate matter into and beyond the GORE-TEX® membrane, which can cause fiberglass fiber breakage due to wear of abrasive dust, and (4) excellent bag-to-cage fit and bag design, which minimized flex fatigue of fiberglass fibers against cage wires, while allowing adequate bag movement on the cage for effective cleaning [187].

However, premature failure of membrane degrades the filtration performance. Although highly efficient, the gossamer-like surface is rather delicate and care must be exercised when handling filter sleeves produced from such materials. However, PTFE filter membranes can also be located within rather than on top of the textile support [188]. The fine surface structure of all membranes implies the need for significant pressure drops across the medium in order to achieve adequate fluid fluxes. One of the most important characteristics of membrane filtration is durability. When the filter bags are subjected to continuous stresses and flexing, particularly during cleaning pulses, the membrane's bond may get broken causing fibril structures and crack development. As a result, filter performance degrades and causes increased emissions, higher pressure drop, increased dust-to-fiber mechanical wear, and ultimately, shorter overall bag life. The extent of severity of these cracks depends upon a number of factors, including frequency of cleaning pulses and amount of pulse pressure used. While using membrane filter, the pulse pressure and pulse-on-time should not exceed 550 kPa and 350 ms, respectively [90,156,189].

Through innovation, PTFE membrane is re-engineered to achieve a high level of strength, yet providing a higher airflow while maintaining the filtration efficiency. To make the membrane stronger, the typical response was to make the membrane thicker, but this also reduced permeability. It was, therefore, challenging to make the membrane stronger, yet finer. Due to the unique structure of new membrane, it can have approximately four times the strength of its predecessor at a particular level of permeability (Figure 32). The new membrane possesses reduced cracking and better resistance to damage from penetrating particles. Once some of the Gore-Tex intellectual property expired, companies such as Tetratex (now part of Donaldson) and BHA Technologies (now part of GE), Mikrotex®expanded PTFE Membrane [108] entered the ePTFE market and went on to

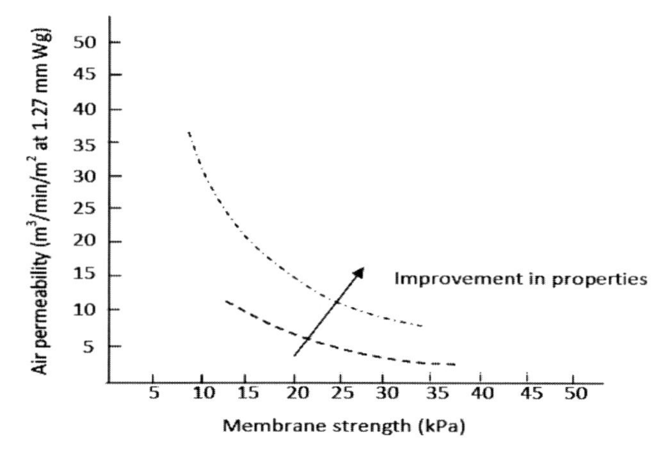

Figure 32. PTFE membrane strength vs. air permeability characteristics.

further develop the technology. Tetratex ePTFE membranes are manufactured from 100% PTFE resin by a unique patented process of Donaldson Company Inc. and are marketed by Donaldson [190].

In general, various ways of polymeric membrane manufacturing are sintering, stretching, track etching, and solvent casting/phase inversion. There is a growing interest in the use of inorganic membranes like ceramic because of their robustness, tolerance to extreme conditions of operation, such as higher temperature and aggressive chemicals, and the resultant long life, which offsets their higher initial cost as compared with polymeric membranes [129]. The preparation and performance of ceramic membrane filters made from fly ash from coal-fired power stations were investigated. The porous and crack-free ceramic composite membrane was prepared by dip-coating a stainless steel mesh in slurry of fly ash. The median pore size of the tested membrane was 2.3 μm and the differential pressure gradient was 0.9 kPa L^{-1} min^{-1}, which is comparable to fabric filters currently in use in some New South Wales power stations. The material is reported to exhibit great potential for filtration and microfiltration of high-temperature gas-particulate streams. As the membrane consists of metal and ceramic composities, it shows good flexibility [191]. The evaluation of dead-ended 'honeycomb' ceramic filters for fine particulate and Hg removal in an industrial boiler showed that the fine particulate emissions were greater from the pulse-jet baghouse than from the ceramic membrane filters in a coal-fired industrial boiler [192–194].

A composite laminated product was found to exhibit better filtration performance. The laminate product includes a first layer of porous-expanded polytetrafluoroethylene membrane, and a second layer of woven fabric of polytetrafluoroethylene-containing yarn. The laminate product can be sewn to provide a filter bag for use in pulse-jet filter applications [186]. However, all such membrane applications are relatively expensive, their use is normally restricted to difficult applications, for example, where the dust particles are extremely fine or are of particular hazardous nature, or where the interaction with a surface of this type shows unique advantages with respect to cake release. As in advanced hybrid technology (Section 6.2.), since the burden of particulates on fabric filter is substantially reduced, the limitation of working of membrane fabric with lower air-to-cloth ratio can be greatly improved.

5.4.3. Coated filters

The coating of needle felt fabric surfaces is a little more complex, and sometimes it is difficult to draw the difference between the coated fabric and the bonded media. Coating of filter media can be done by various ways, such as surface coating, steeping, foam coating, and coating with inert powders.

5.4.3.1. Surface coating. Microporous polymer coatings can be applied to the face of nonwoven fabrics, both to achieve finer filtration and improve cake discharge. The coating may be sprayed on the base fabric as a liquid, or laid down as a thin sheet that is then bonded to the fabric, or even pushed into it. Through the coating process, heat, chemical and abrasion resistance, and cake release properties of the filter are improved besides filtration efficiency [195]. Coating forms a quasimembrane structure on the surface of the needle felt.

A coating of polyurethane on polyester or acrylic felts impart hydrophobic and oleophobic characteristics, whereas a chemical treatment of the felt with a resin containing

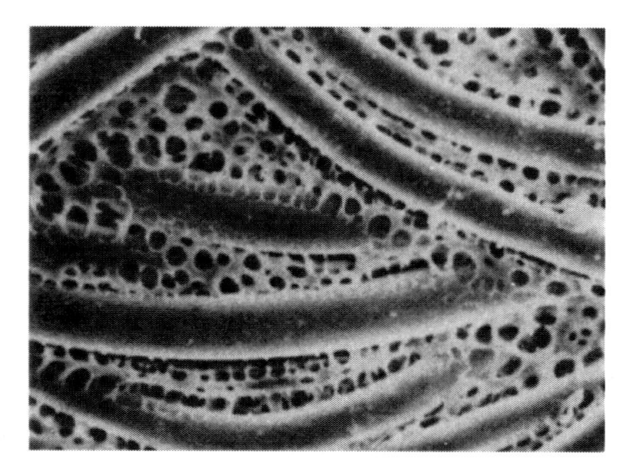

Figure 33. Ravlex coating on spun-bonded polypropylene needle felt.

PTFE on aramide, polyamide, and polyphenylene sulphide (PPS) provide good chemical resistance. Surface coating of Teflon B on glass fiber fabric is common for many boiler applications [133]. For achieving superior filtration performance, a range of coated material was developed by Ravensworth Ltd. (in the late 1980s) with the trade name of Ravlex. In one of the processes, Ravlex coatings made out of tetrafluoroethylene terpolymers were applied in liquid form to provide a very robust coating with 5–8 μm pores. The scanning electron micrograph in Figure 33 shows a Ravlex coating on a nonwoven fabric. However, in general, the process comprises three main product groups: Ravlex MX, Ravlex PPC, and Ravlex YP. Ravlex MX is a PTFE microporous-embedded coating that is applied to the filtration surface of the carrier needle felt. Ravlex MX is generally used for the filtration of fine particles in the chemical, pharmaceutical, and food industries. The Ravlex PPC coating is also formulated from PTFE and it possesses a larger pore size resulting in higher permeability than the Ravlex MX. This makes it suitable for filtering agglomeration or sticky dusts. Ravlex YP is formulated from polyurethane (PU) and has a pore structure similar to Ravlex PPC. Yet another product group recently reported is Ravlex CR formulated from PTFE coating. All these coatings usually offer high-abrasion resistance, effective cake release, higher air permeability, reduces operational pressure-drop (resistance to blinding), reduces frequency of cleaning, and extends operational life of filter media [196–199].

MicroWeb 2000 and MicroWeb II media, introduced by Webron, comprise polytetrafluoroehylene (PTFE) and acrylic coatings, respectively, on a polyester needle felt. Both are designed to offer relatively high permeability. The company also offers chemical treatments, under the brand name 'Supaweb', which are applied and thermally bonded to the base substrate. Each treatment provides particular additional property to the felt, e.g. improved cake release, hydrophobicity, and improved chemical resistance. The company Fratelli Testori produces a number of treatment processes including *Novates*, a coating of polyurethane (PU) on polyester or acrylic felts, possessing hydrophobic and oleophobic characteristics and *Mantes*, a resin chemical treatment containing PTFE that is applied to acrylics and high-temperature fibers to give a good chemical resistance. Madison Filter has developed Tuf-tex^{TM}coatings for polypropylene, polyamide, and polyester substrates. These are thermosetting resins sprayed or knifed onto the surface of the filter fabric to provide good abrasion resistance along with improved dimensional stability [297].

5.4.3.2. Steeping. In an application, after steeping the fabric in a chemical solution containing PTFE and fluoride resins at high concentrations, the fabric is dried and heated to fix the fluorides on the fibers. It is used on polyester or acrylic fiber fabrics to impart good cake release property and protect from chemical activity [90]. In the case of treatment to glass fiber fabric, the fabric is immersed in an emulsion of water, silicon, silicon graphite or colloidal graphite, or fluorocarbon compounds, then squeezed through a roller to remove excess of liquid, and finally dried [13].

5.4.3.3. Foam coating. Foam treatment is achieved by the following steps:

1. Mechanically generating a low-density latex foam.
2. Applying this foam to the fabric by knife over roller or knife over air technique.
3. Drying the foam at a modest temperature.
4. Crushing the foam to produce an open-cell structure.
5. Curing the foam at a higher temperature to crosslink the chemical structure.

The principle ingredient of the treatment is usually aqueous-based acrylic latex; the precise formulation may comprise a variety of chemical agents to ensure the production of a fine, regular, and stable pore structure and to provide specific characteristics such as antistatic or hydrophobic properties. Polymer selection for coating is governed primarily by surface chemistry factors and there must be chemical affinity between the contiguous materials involved in the finished products. The electron photomicrograph of a foam layer is shown in the Figure 34.

Acrylic foam-coated needle felts produced in this manner are capable of continuous operation at a temperature of approximately 120°C. However, they are not normally resistant to hydrolytic conditions, these leading to collapse of the structure and hence premature pressurization. The actual density of the foam as applied to the material is also critical to a successful application. High density leads to excessive wetting of the substrate and resulting in an unacceptable air permeability and too low density leading to inadequate penetration, poor mechanical bonding, and hence the risk of delamination.

5.4.3.4. Use of inert powders. Chemically inert additives, light density powder that is injected into the baghouse establish a uniform and porous dust cake on the filter bags with enhanced operational efficiency and providing bag protection from moisture, particulate bleed through, hydrocarbon carryover, bag blinding, oil, and tacky or viscous contaminants. Powders are chemically inert additives, which result in uniform and smaller sized pores on the fabric surface. In a report [108], improved filtration performance was claimed to be achieved by injecting chemically inert, light density powder Opti-Coat^{TM} into the baghouse. This powder is extremely light in weight and will not penetrate through the fabric filter even with higher velocities, and consequently does not 'blind' the fabric surface. Chemically inert powders like this can be used on an ongoing basis of application for chemical attack protection.

By pre-coating this powder onto the new fabric filters, a protection coat is developed on the filter surfaces. Powder coating can be done with the help of calendering. The powder, which is scattered onto the surface of nonwoven fabrics, is of relatively lower melting point than that of nonwoven fabric. During calendering, the powder will melt with fibers, and form a layer of smaller pores.

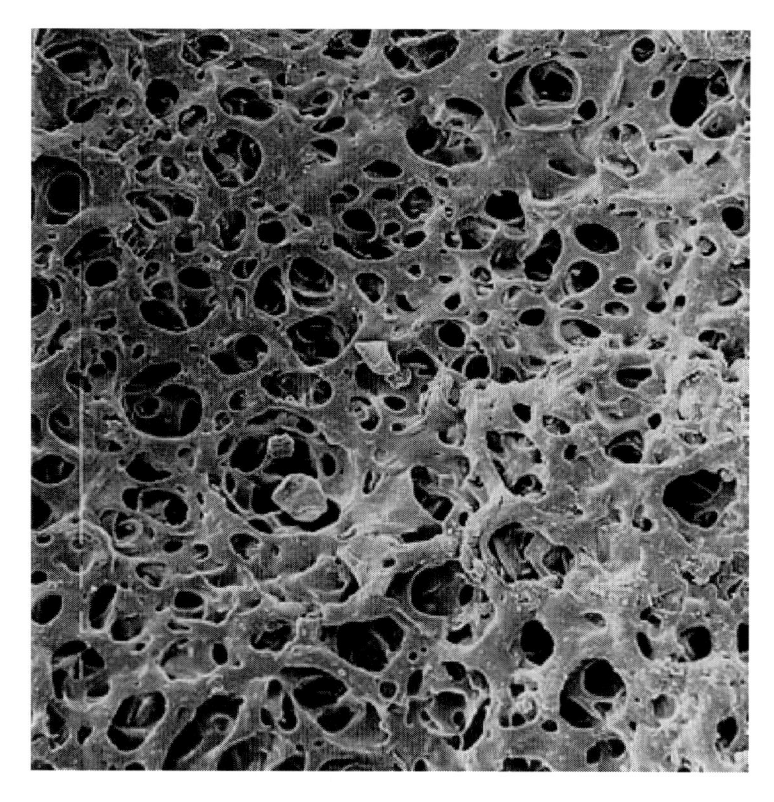

Figure 34. Electron photomicrograph of a foam layer on a nonwoven fabric.

5.5. Layered structure

Filter materials with layered structure may influence the performance of the filter fabrics. For enhanced surface filtration, layered filter fabric can be made by incorporating finer fibers into the upstream side of filter fabric [200,201]. Figure 35 shows the impact of finer fiber content on the upstream side of a layered needle-punched nonwoven fabric on filtration efficiency and pressure drop. It was observed [200] that a certain percentage of finer fibers in the upstream side of needle-punched fabric could provide improved filtration performance. Fabric consisting of only finer fibers exhibits higher efficiency at the cost of higher pressure drop. This is due to greater specific surface area of finer fibers, higher level of consolidation (fiber flexibility coupled with barb needle to take up more fiber during downward stroke in the process), and feasibility of achieving a smaller size of theoretically defined pores. Since the upstream side is mainly responsible for filtration, the properties of finer fiber can be utilized only at the upstream side; the role of coarser fiber downstream provides the strength to the material without much enhancing of the pressure drop.

Callé et al. [202] investigated nonwoven filters with different surface properties on filtration performance. The study has shown that treating the surface with thermobonding or covering the surface with a fine layer of submicronic fibers promotes detachment of the dust cake and prevents particle penetration inside the initial porous media. During filtration as cycles progress, the starting pressure drop (preceding cleaning) increases sharply for filter without any surface treatment. This increase is attenuated when the media has undergone surface treatment [202].

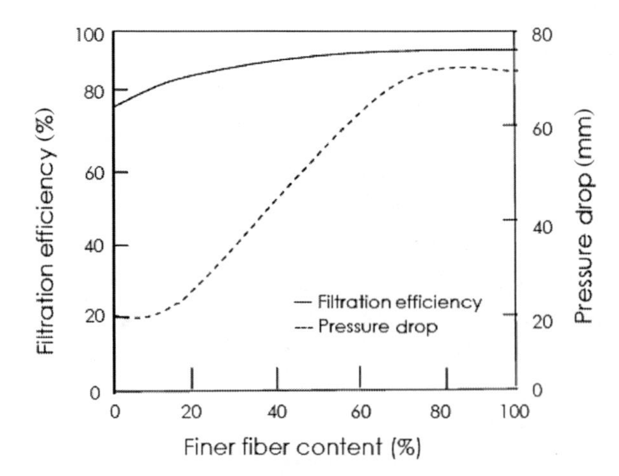

Figure 35. Effect of finer fiber content on filtration performance [200]. (Reprinted from V.K. Kothari, A. Mukhopadhyay and S.N. Pandey, *Filtration characteristics of layered needle punched nonwoven fabrics*, Melliand Textilberichte, 74 (1993), pp. 386–389, with permission of Melliand Textilberichte.)

To achieve improved filtration performance, a sandwich structure was fabricated with two nonwoven fabrics and a polyester grid (C) laminated together to form nonwoven composite for use as filter bags [203]. One nonwoven component (A) was made up of polyester fibers of various levels of fineness including ultrafine fibers and of low melting-temperature (T_m) kept on the surface layer; the other (B) was made of 2.22 dtex polyester fibers. The nonwoven composites were composed of the above material in an A/B/C/B order to form a sandwich structure with a basic weight of 500 g/m^2. The material was thermal calendered and needle punched to bond the layers in the fabricated composite nonwoven fabrics. The sandwich structure could be molded into various shapes to suit different filtering media.

Filtration performance of P84 needle felt can be improved by using microfiber (1.0 dtex) at the upstream side of felt made out of standard 2.2 dtex fiber. This has the benefit of raising the efficiency, while at the same time preventing the dust penetration into the depth of the filter media, maintaining the pressure drop at a low and stable level with extended bag life in most cases. P84 fibers can also be used either as surface layers on polyester, homopolymer acrylic, m-aramide felts, or as a blending partner for PTFE fibers [171].

Under layered construction, different combinations such as felt-like layered composite of PTFE and glass paper were reported [204]. It was also reported that the high efficiency nano-microfiber composite filters could be used in industrial filtration. A thin web made out of interconnected fine fibers can have very small pore size [205], which is useful for effective filtration. The nanofiber-based filtering media, made up of fibers of diameter ranging from 100 to 1000 nm, can be conveniently produced by electrospinning technique. Nanofibers can provide an improvement in filter efficiency, without substantial increase in pressure drop. They have proven to enhance the life of filters in pulse-clean cartridge applications for dust collection [206]. A filter bag medium composed of 70% regular and 30% low-melting polymer in polyester felts as base material and a prepared polyamide nanosized web at the upstream side of the fabric was also reported [207]. The filter medium was overlapped and thermal bonded by a thermal roll calender. In the evaluation of filtration test, the filter medium with nanosized web showed more stable filtration behavior and improved collection efficiency. Figure 36 shows SEM images of the filter medium with

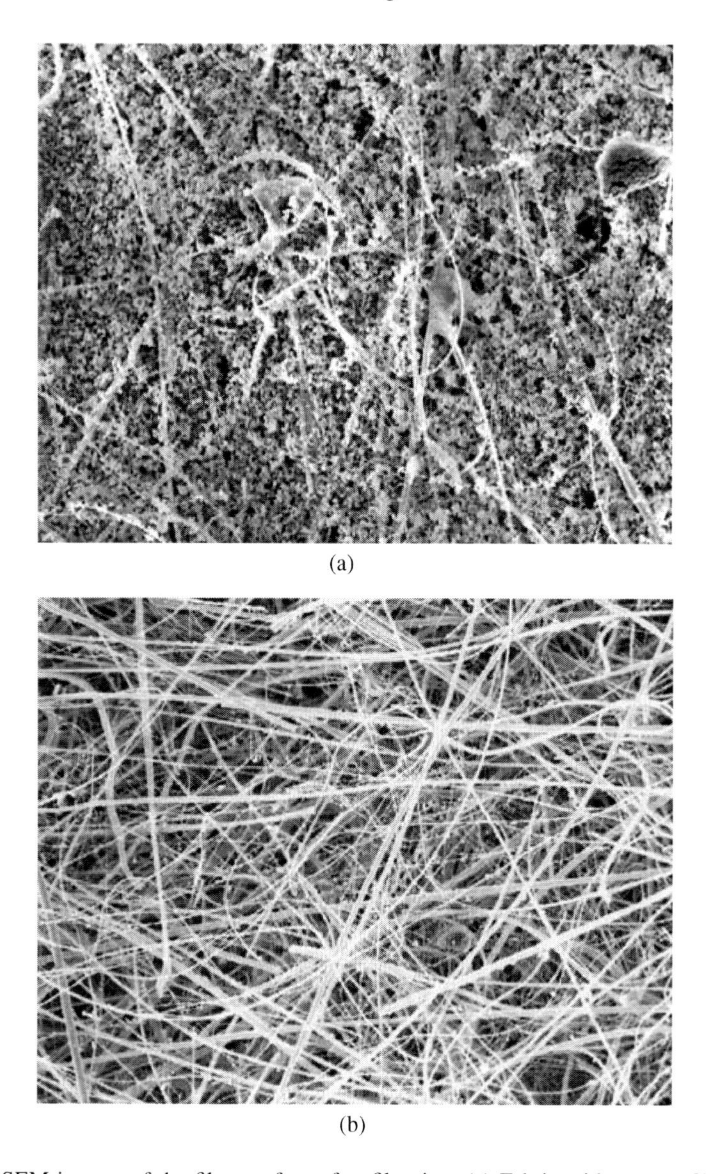

(a)

(b)

Figure 36. SEM images of the filter surface after filtration. (a) Fabric without nanofiber web; (b) Fabric with nanofiber web (× 500) [207]. (Reprinted from S.Y. Yeo, D.Y. Lim, S.W. Byun, J.H. Kim and S.H. Jeong, *Design of filter bag media with high collection efficiency*, J. Material Science, 42 (2007), pp. 8040–8046. Reprinted with kind permission of Springer Science and Business Media, Heidelberg.)

nanosized web after filtration. Very fine dust is found to be captured effectively by the medium. Presence of nanofiber web in fabric also shows that the amount of dust left in fabric (after filtration test) is much lower than the fabric without a nanofiber web [207]. Use of nanofiber/celulose composite was also reported for pulse-jet filter cartridge in the industry [206]. In a patent disclosure, use of fine fiber (microfiber/nanofiber) structure on the upstream surface of the spun-bonded filtration substrate was mentioned for achieving high-filtration efficiency [208].

5.6. Filter element and its fabrication

In fabric filter, the fabric is given a shape or form to take up the filtration process. Depending on the type of the filtration or the design of the filter, the fabric is fabricated in the form of cylindrical filter element or cassette/envelope. The most commonly and widely used form is a cylindrical one [13]. The filter element fabrication is of vital importance and therefore the following points are considered during selection:

1. Size and shape of the filter element.
2. Required slack (the ratio of circumference of filter element to the circumference of support cage. The normal slack is 1.05).
3. Component parts like cuffs, anticollapse rings, etc.
4. Seam construction.
5. Type and pattern of sewing stitches, fabric tension, type of stitching needle and its size.

6. Separation of difficult and fine particulates

Separation of difficult and fine particles requires extra attention and additional arrangement. Apart from the selection of fabric, designing of the whole filter assembly may be quite different depending on the type of problem.

6.1. Separation of difficult particulates

The conditions in the process industry are extremely diverse and often demanding with respect to gas atmosphere and particle properties. Dust particles are difficult to separate or to handle if the following conditions exist:

- The flux density of the particles to the collecting surface is low leading to too small collection efficiency.
- The particles cannot be removed from the collecting surface or from the separator due to their strong adhesion forces.
- Light and fluffy dust, and particles with poor flowabilty.
- The particles have extreme mechanical or chemical properties, e.g. abrasive particles, very hard or very soft particles, particles with high reactivity.
- Hot particles.
- Particles causing blinding of pores.
- Nanoparticles with adverse gas properties.
- Mixtures of solid and liquid particles in the aerosol.
- The particles do not adhere to the collecting surface due to low sticking efficiency.
- The particles cause fire hazards/explosion.

Particles with one or more of these properties occur in a wide variety of applications, e.g. these range from the flame synthesis of nanoparticles in the gas phase to condensation aerosols, which occur in the wood industry, in soil remediation, in waste-recycling processes, or in municipal waste incinerators. Technological limitations of controlling difficult particles and some designing aspects for overcoming these problems are mentioned below.

6.1.1. Light and fluffy dust and particles with poor flowability

Light and fluffy dust often gives re-entrainment problems in the collector. Collected dust on collecting plate in an electrostatic precipitator or on bags of fabric filter is cleaned for the dust cakes to fall in the hopper. Light dust that has not been able to form a cake is disintegrated during cleaning and carried away by the incoming gas and thereby is difficult to collect. It is therefore necessary to be careful in the selection of the velocity, which has to be as low as possible.

Flowability is an important criterion for the transport of dust out of the separator. For fibrous particles, extremely high-separation efficiencies may be needed due to the high impact of some type of fibers on human health (e.g. asbestos). Fibrous dust can cause problems due to clogging, particularly when the distance between the individual filter bags is too small [209]. Very fine dust or fibrous dust can also have the tendency of building up in the ducts, within dust collector and often leading to clogging of duct, dust hopper, etc. [13]. Proper designing of filter unit as well as selection of operating parameters can solve the problem by a large extent.

6.1.2. Particles with strong adhesion

If the particles are sintering, e.g. reactive powder, paint particles, or particles with chlorine components in high-temperature filtration applications, caking and/or problems due to dust cake removal become pronounced. Certain dust, like coal tar, is inherently glutinous and tends to deposit and build up in the collector, which is difficult to remove. Furthermore, another more difficult condition in dust collection arises from the presence of moisture in the gas stream or if the dust is of a sticky nature from previous processing. The dusts which absorb moisture are called hygroscopic. Dusts containing chlorides, organic dusts with large surfaces and fibers, aluminium fluorides, certain hydroxides, calcium oxide, calcium chloride, and most synthetic fertilizers are hygroscopic in nature. These type of dusts pose problems to fabric filters because of frequent clogging of the fabric. The problem is predominantly serious at the starting and stopping of the filter as well as when the operating temperature is below the dew point temperature. In such conditions, condensation occurs, which leads to clogging of fabric pores. This situation will be aggravated if the fabric is subsequently allowed to dry out, resulting in the formation of nodes or agglomerations of dust particles and leading to an increased weight of dust cake and eventually a critical blinding situation. General recommendation to avoid the problem is to run the fabric filter with hot air after shutdown and before startup. In addition, insulation of the casing and hopper may be necessary. It is also advisable for the fabric to be subjected to special hydrophobic or oleaphobic treatments as part of the finishing process [13,90].

6.1.3. Abrasive particles

Abrasive particles are characterized by specific shapes like sharp edges and high values of their hardness. Abrasive particles can be harmful in all applications where these particles have a high relative velocity (e.g. in poorly designed baghouses). Dusts from grinding or polishing applications or ceramic particles like SiC, dust like alumina, fly ash, clinker, etc. are highly abrasive and create severe erosion problems if not properly taken care of during the selection of dust collectors. Controlling the velocity of gas is important since high velocity leads to greater penetration of particles. Trapped abrasive particles usually cut the fibers in contact during the flexing of the fabric at the time of the cleaning process. While selecting the bag filter, care should be taken to choose the type of fabric material and its finish. Abrasive dust also poses a problem to dust-conveying equipment like screw conveyor, rotary airlock, etc. Replaceable liner of special material in rotary airlock and hard facing with weld deposits on screw flights is resorted to [13].

6.1.4. Hot particles

The presence of very hot particles generated from combustion, drying, or other processes can be carried with the gas stream into the filtration compartment where they can cause a hole in the fabric and present a serious risk of fire. High-temperature filter elements can be distinguished into glass, ceramic, and metal filter media made of fibers or granules. For ceramic and metal materials, different structures in the form of fiber fleeces, woven tissues, and sintered granules are available. Maximum operational temperature for metal elements, depending on the steel grade, ranges up to 600°C. With ceramic elements, the operational temperature range can exceed 850°C. Sometimes, an inert protective pre-coat layer on the thermal resistant filter surface can solve the problem [209]. In certain conditions, even ostensibly nonflammable polyaramid fibers have been found to ignite. Consequently, if adequate particle screening is not provided, the fabric may require a special flame-retardant treatment [90].

6.1.5. Particles causing blinding of pores

In case of the bag filter, blinding tendency by the particles results in high differential pressure across the bag in a short period of time and therefore it is detrimental for smooth running of the filter unit. For example, in a milk powder plant, blinding and bag damage in the main washable baghouses can cause interruptions to the dryer operation. Further, as the baghouse air-to-cloth ratio increases, penetration of powder into the bag fabric during operation also increases [57–59]. These particles imbedded in the pores of the filter medium cannot be removed by reverse flow or pulse cleaning [57,60–62], consequently resulting in higher differential pressures. Baghouse operating costs increase when operating at higher differential pressures, firstly because the energy costs for running the fans that suck or blow air through the baghouse are higher [63–66] and secondly, because bag replacement costs increase as bag life is shortened by operating at higher differential pressures. Appropriate air-to-cloth ratio is required for avoiding/reducing blinding tendency [164]. For some active dust, suitable filter fabric design is necessary to avoid blinding of filter fabric.

6.1.6. Nanoparticles with adverse gas properties

In various applications, nano-sized particles have to be removed from highly concentrated acidic gases. Process gases from flame synthesis of TiO_2 or SiO_2, for instance, contain significant portion of chlorine. In other processes (e.g. soil remediation), considerable amounts of condensables, such as tar components, may cause problems due to irreversible plugging of the filter elements. To avoid any contamination of the collected powders, careful selection of filter media and materials of construction is required [71]. Most of the commercially available high-performance filter media show excellent separation characteristics, and, to some extent, a reliable long-term filtration behavior. Excellent separation efficiencies were also found for nanoparticles like fumed silica. Specific cake resistance of fumed silica was higher than those obtained for quartz by about three orders of magnitude [210].

A bag filter can be designed for the separation of TiO_2 nanoparticles after a steam-jet mill operated at approximately 130°C in a gas atmosphere of almost 100% water vapor. The raw gas inlet is from the top through the clean gas chamber to the bottom so that an almost ideal down flow configuration is realized. The baghouse is heated and insulated so that condensation of water vapor is avoided. During online regeneration of the filter medium, the pulse gas has to be heated in order to avoid condensation in the pulse-jet [209]. It is also important to add that hybrid filters combining electrostatic precipitator-fabric filters and wet scrubber-fabric filters are effective in the removal of nanoparticles.

6.1.7. Separation of mixture of solid and liquid particles

There are a variety of solutions to separate mixtures of solid and liquid particles with difficult dust properties. The principal ways to reach this goal are as follows [211]:

- Enhance efficiency of transport mechanisms by introducing additional forces on the particles (e.g. electrocyclone).
- Change the particle properties so that they can be collected easily (e.g. heterogeneous condensation in scrubbers).
- Avoid clogging of filter media by means of pre-coating (e.g. tar collection).
- In case of sintering, changes in the process technology are required.

6.1.8. The particles causing fire hazards/explosion

In general, fire and/or explosion may happen due to the presence of (a) a substance or material which may ignite or combust, i.e. combustible gas or vapor, combustible dust, mist or combustible liquid, or combination thereof, (b) oxygen that supports the combustion, and (c) sources of ignition such as mechanically produced sparks (as in the case of process of grinding, shot blasting, etc.), glowing particles or flames (as in the case of a metal-melting furnace), combustible gases like carbon monoxide, heat generated due to exothermic reaction, generation of static electricity during fabric filtration [52]. It may be added that the explosion problem is more critical in fabric filters as bag replacement is inevitable in the case of a burnt/damaged bag due to explosion.

In many cases, a dust collector can handle gases or dusts, which are explosive and/or inflammable in nature under certain conditions. For example, metal powder in suspension in air or fine pulverized coal dust in coal mill exhaust gas are potential hazards for explosion. In powder form, metals like iron or aluminium etc. and nonmetals like wood, plastic, coal, rubber, flour, etc., may cause serious explosions and fire hazards unless proper precautions are taken. It is not only the property of the material that decides the explosive nature of the material but the other factors like the fineness, content of the powder dispersed in the gas/air, rate of rise of pressure, etc. that matter most. This is the reason that a block of iron will not explode or ignite but iron powder in oxide form may propagate fire very fast in the presence of spark and oxygen.

Normally the control equipment operates under negative pressure but when explosion takes place, very high positive pressure is developed. The severity of explosion depends on how fast the pressure rise takes place. Rate of pressure rise is therefore an important phenomenon and depends on the dust hazard class and dust concentration. Rate of pressure rise is defined as highest (maximum) value of the pressure rise per unit time during the explosion of a given dust in a closed vessel at optimum concentration. Class of material is designated as 1, 2, and 3 depending on the severity of explosivity of the dust in ascending order; 3 being the highest risk of hazard and having higher/faster rate of pressure rise [52]. For prevention against fire hazards and explosions, selection of filter media and safety features adopted in filter units play a very important role. Safety features in filter units for explosive and/or inflammable dust particles are discussed in detail in a separate section. Herein, selection and designing aspects of various filter fabrics are discussed.

The natural tendency of the fabric is to acquire a static charge because of friction of fibers and dust at the time of filtration while the dust is passing through the pores. The particles that are conveyed to the collector may also possess an electrostatic charge [167], either pre-applied or acquired en route that, if carried into the collection compartment, could accumulate with potentially explosive consequences, e.g. a case involving white sugar dust-handling systems [212]. It is important to note that pulse-jet dust collectors posses an inherent advantage in mitigating explosion risk due to static charge since their design uses the reverse flow of air. This reverse flow of air (that is not ionized) through the dust depletes all static charge from the dust particles. The cleaning system is thus constantly cleaning to prevent any charges from building up [149]. However, if charges build up to a level, it can trigger off sparks. If the gas is inflammable or the dust is explosive, sparks can lead to fire. The fire due to incomplete combustion produces carbon monoxide whose level, reaching a dangerous level, finally causes explosion [52].

Extreme care must be taken in selecting and designing dust collector-handling explosive or inflammable dust since explosion causes damages to material, personnel, and properties. Explosion occurs at specific dust concentration and temperature, generally in the presence of spark or flame. However, there are dusts like pulverized coal that ignite by themselves when accumulated. Sugar, flour, peat powder, pulverized coal, aluminium, magnesium, sulfur, wood dust, malt, plastic, zirconium, etc. are considered to be very explosive dusts [13].

To guard against the above-mentioned risk, the system must be well earthed, which is only possible if the fabric of the filter bags has a sufficiently high electrical conductivity. By contrast, with this requirement, the synthetic needle felts have a high electrical resistance, and are therefore very susceptible to get highly charged with static electricity [129]. Bringing down the specific resistance to 1×10^4 Ohm cm is good enough to make the fabric sufficiently conductive to discharge the charges to earth. With synthetic fibers, most of which are nonconductors, early efforts were made to include copper or stainless steel wire into the fabric for draining the charge. The limitations of this design soon became evident when it was seen that the metal wires are not sufficient to dissipate the charge accumulated inbetween the successive wires. On the other hand, increasing the number of wires posed the problem of reduced filtering area [13].

As static electricity is essentially a surface effect, the filter fabric should be given antistatic properties by a special surface treatment or through the inclusion of antistatic fibers such as stainless steel or carbon-coated polyester (epitropic). Normally 3–5%, by weight of the conductive material in blending, gives effective conductivity to dissipate the charge efficiently. Cotton, being conductive to electric charge, remained a natural choice for a long time before the introduction of suitable synthetic fabrics for an effective solution to the explosion. By nature, acrylic fiber is antistatic but has a poor tensile strength. Therefore, the fiber can be blended with polyester to achieve adequate strength. Chemical treatment has the disadvantage that it is not durable, since the coating is likely to abrade and disintegrate in use. On the other hand, the inclusion of conductive fibers provides permanent protection. Blending of carbon fibers with other fibers during manufacturing of felt, followed by additional emulsion treatment, was reported to be very effective [13].

It was also claimed [213] that by constructing the filter medium with a blend of fibers of widely contrasting triboelectric properties, superior collection efficiency can be obtained. It is further claimed that by virtue of this enhanced efficiency a more open structure can be used with consequent advantages with respect to the reduced power consumption required to pull the dust-laden air through the collector. However, although this effect has been used to some advantage in clean air room filtration applications, considerably more research is necessary if the triboelectric effects in industrial dust collection are to be fully understood and exploited.

In a patent disclosure [214], a nonwoven textile material containing an electrically conductive polytetrafluoroethylene fiber was reported to be useful in dissipating static electric charges. DuPont marketed Bekaert's Bekinox and Epitropic as conductive fibers [129]. The former one is of a special grade of stainless steel that was of high purity, so as to avoid the risk of inclusions within the very fine diameter fibers. Epitropic fibers are primarily polyester, with an outer sheath of polyester isophthalate copolymer, which is impregnated with particles of carbon black. The sheath has a melting point of 35°C lower than the core; it can be softened by controlled heating so that the carbon particles embedded in it become an integral part of the fiber surface. The electrical conductivity of these fibers is 50 times higher than that of stainless steel; this, combined with their significantly lower density, was claimed to have significant cost advantages for lower temperature applications suited to polyester [129].

6.2. Hybrid technology

The primary technologies for the state-of-the-art particulate control are fabric filters (baghouses) and electrostatic precipitators (ESPs). However, each of these has limitations that prevent it from achieving ultrahigh collection of fine particulate matter. Among the major shortcomings of ESP performance are its dependency on resistivity and the particle size of the dust. In case of ESPs, the fractional penetration of 0.1 to 1.0-μm particles is typically of an order of magnitude greater than for 10-μm particles, i.e. electrostatic deposition is effective for relatively large particles, but it is quite ineffective for the ultrafine ones because their charging probability in the corona field is too low. Therefore, a situation exists where the particles that are of greatest health concern are collected with the lowest efficiency. Fabric filters are currently considered the best available control technology for fine particles, but they also have some limitations. Collecting very small particles through changing the design of fabric filter is possible at low filtration velocity to keep the operating pressure drop within limit. On the other hand, at low filtration velocity fabric filters are large (as total quality of incoming gas volume remains the same), require significant space, are costly to build, and unattractive as replacements for the existing precipitators. Reducing their size by increasing the filtration velocity across the filter bags usually results in an unacceptably high pressure drop and reduced collection efficiency. In addition, many fabrics cannot withstand the rigors of high SO_3 flue gases, which are typical for bituminous fuels.

In an effort to overcome the above-mentioned deficiencies (i.e. to increase collection efficiencies with a moderate increase in pressure drop), a number of efforts have been made, and more are underway to develop hybrid particle collectors. These hybrid devices attempt to integrate electrostatic and barrier filtration into a single device or system, and efficient filtration of ultrafine particles is predominantly done in two mechanisms – electrostatic and diffusional deposition. Specific anticipated benefits of this approach are as follows:

- Reduction of pressure drop and penetration by almost an order of magnitude compared to standard fabric filters;
- Provides ultrahigh collection efficiency, even at high air-to-cloth ratios;
- Virtually eliminating the problem of re-entrainment and recollection of dust in conventional pulse-jet baghouses caused by the close bag spacing and the effect of cleaning one row of bags at a time;
- Overcomes the problem of chemical attack on bags;
- Requires significantly less total collection area than conventional ESPs or baghouses;
- Is suitable for new installations or as a retrofit replacement technology.

Electrostatics can be added to fibrous filters in three ways:

1. *Filtration in an electric field without particle charging:* The field can be produced between a pair of electrodes (e.g. of cylindrical configuration) with a fibrous filter placed between these electrodes [147]. Every second wire is maintained at a high

potential, while the others are grounded. The best filtration results were obtained for the electrodes located at the dusty side of the filter [215]. The electrodes can also be excited by high-voltage air current, but this mode has only a negligible effect on the collection efficiency [216].

2. *Filtration of charged particles without electric field:* The particles are charged in a corona discharge, in a stage before the filter, and removed in a traditional bag filter with no electric field applied [217].
3. *Filtration of charged particles in an electric field:* The particles can be charged before their filtration in a corona-free electric field [218–221], or simultaneously charged and precipitated in the same stage [167,222–224]. The concentrically arranged corona wire charges the particles in the lower part of the filter. Larger particles are collected on a metal tube, which in fact is a tubular ESP, while the particles of lower mobility flow upwards to the bag filter where they are deposited.

However, in all the above cases, the filtration system should be carefully monitored under controlled conditions. In practice, the hybrid particulate collection concepts include

- electrostatically enhanced fabric filter;
- compact hybrid particulate collector;
- advanced hybrid particulate collector; and
- multi-stage collector (MSC).

6.2.1. Electrostatically enhanced fabric filter

The collection efficiency of fibrous filters can be increased by electrical energization, particularly for the particles of submicron size range. The penetration of such particles can be decreased from a typical value of 90% down to about 10% or even lower. The pressure drop is also lower in this type of filter due to changed structure of deposited dust under the action of an electric field. Due to lower pressure drop, the gas flow rate can be increased without changing the collection efficiency, allowing the reduction in the number of bags used for filtration. Regeneration of electrostatically assisted bag filters is efficient due to the enhanced surface filtration [225].

There are a large number of parameters influencing the performance of pre-charging, which are as follows:

- Initial porosity of the dust cake;
- Electrical field in the dust cake;
- Particle size and charge distribution;
- Pre-filtering effects in the pre-charger.

The increased porosity of the dust cake is due to the formation of longer dendrites when charged particles deposit on a single fiber or in the dust cake. As the particles reach the filter surface, they build up an electric field. The resulting field depends on the particle charge and the leakage of current from the dust cake. This, in turn, depends on the effective electrical conductivity of the dust cake. The presence of an electric field at the filter surface shows a correlation with the external field case, which is an important necessity for an increased porosity [226].

In another study, pre-charging of particles entering a fabric filter were shown to have a significant effect on both the collection efficiency and pressure drop during filtration of the fly ashes. Collection efficiency can be enhanced significantly by the charging of the fly ash particles. The beneficial effect of the particle pre-charging depends strongly on the gas velocity in the system, the dust concentration, and the triboelectric charge on the fly ash. A high charging of particles generally leads to a pressure drop reduction and an increase of collection efficiency. The pressure loss of the dust cake as well as the cycle time between back-pulse cleaning was improved under certain conditions. At room temperature, the pre-charger showed a lowering effect on the filter pressure drop at low filtration velocities. This beneficial effect disappeared at higher filtration velocities. At higher temperatures, when the fly ash revealed high electrical resistivity, the detrimental effects of back-corona decreased the pre-charger performance. Intermittent energization of the HV-power supply was applied to reduce back-corona effects. However, with the system used in this investigation, the corona voltage did not drop below corona onset in the pauses between the voltage peaks. Therefore, this type of intermittent energization showed no improvement of the pre-charger efficiency [217].

In another study, dust particles charged in a pre-charger inflowing tangentially into the system vessel were collected on the vessel's inside-wall by centrifugal and electrostatic forces; and the collection was made by the electrostatic force and filtration mechanism in the fabric filter media [227,228]. The reduction in pressure drop is the cumulative effect of decrease in dust loading on the fabric filter, and the formation of a dendrite structure caused by the inter-repulse force due to the same polarity of particles deposited on the fabric filter [229,230].

In order to overcome the low collection efficiency for submicron particles and high pressure drop [231], the characteristics of an electrostatic cyclone/bag filter with inlet types (upper and bottom inlet) were studied. The schematic view of the arrangement is shown in Figure 37. It was found that the electrostatic cyclone/bag filter represented an increment of over 5% for the collection efficiency of submicron particles (around 1 μm) in comparison to the general fabric filter [231].

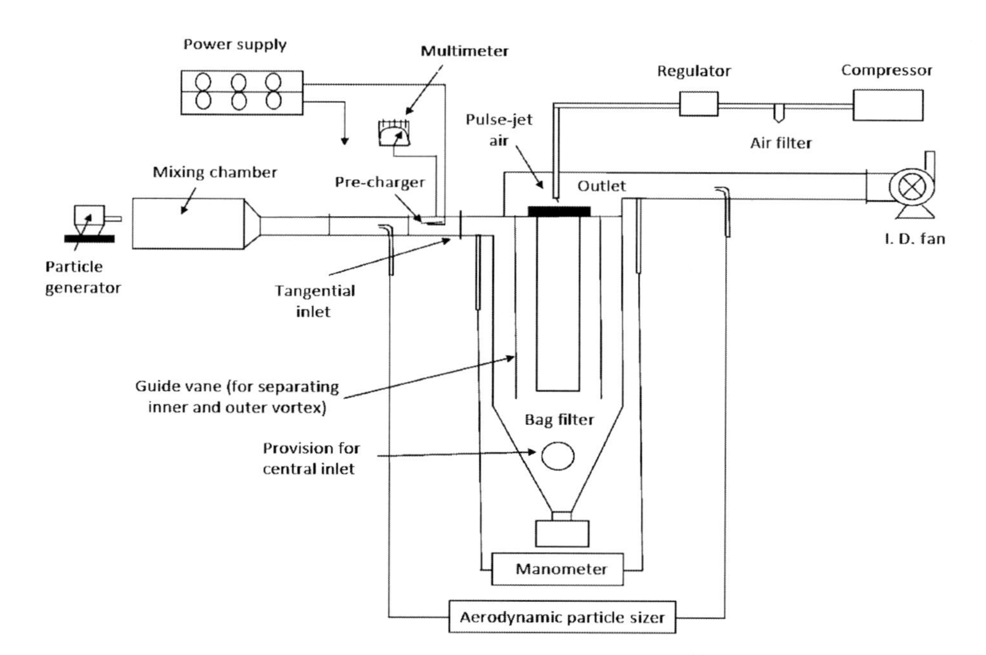

Figure 37. Combination of electrostatic cyclone bag filter [231]. (Reprinted from S.-J. Yoa, Y.-S. Cho, Y.-S. Choi and J.-H. Baek, *Characteristics of electrostatic cyclone/bag filter with inlet types (Lab and Pilot scale)*, Korean J. Chemical Engineering, 18 (2001), pp. 539–546. Reprinted with kind permission of Springer Science and Business Media, Heidelberg.)

Yet, in another study, a hybrid dust collector [232] combining electrostatic charging with fabric filtration was developed and its performance characteristics were evaluated. The nonwoven polyester fabric coated with porous acrylic surface was used for the investigation. Charged particles build porous dendrite structures on the filter surface by electrostatic attraction, increasing the collection efficiency of dust particles and reducing the pressure drop through the deposited dust layer and filter media. For the dust-charging condition, the average cleaning interval was prolonged by about twice as much and the residual pressure drop after cleaning was reduced by 24% from that for the uncharged condition. The cleaning performance of file dust layer is improved because the dendrite-structured dust layer can be removed more easily by pulse-jet cleaning flow. The results of the experiment showed a reduction of fine particle emission by 37%, and 13% energy saving by pre-charging dust particles before filtration [232].

6.2.2. Compact hybrid particulate collector

In COHPAC, a high ratio pulse-jet fabric filter collector (baghouse) is installed in series after an existing ESP. COHPAC was originally described by the United States Patents as COHPAC I and COHPAC II [233,234] – both got expired recently. The US Patent 5,158,580, is usually referred to as COHPAC II (FF and ESP integrated), whereas the US Patent 5,024,681, referred to as COHPAC I, describes a FF in series with ESP. Hamon Research-Cottrell was the first implementer of COHPAC under license to Electric Power Research Institute (EPRI). The claims for both the patents specify a filtration velocity range of 8 to 40 ft/min, which wound up being somewhat impractical (Texas Utilities/Big Brown at ~ 15 ft/min was problematic, Alabama Power/Gaston at ~ 8 ft/min was workable) and subsequent industry application of the ESP/FF in series concept fell outside of the range. However, the better the performance of the remaining fields, the higher the final filtration rate [235–237]. It may be added that both of the original concepts have not been implemented at commercial scale.

A standalone ESP (usually existing) followed by a new FF (usually at 4 to 6 ft/min) is better described as a primary ESP followed by a secondary *polishing fabric filter* (Figure 38a). A fabric filter installed in the rear fields of an ESP casing would be better described as *an integrated ESP with polishing FF* (Figure 38b). This configuration could be either a retrofit into an existing ESP or a new unit. As a retrofit application, the technology is especially appealing for units that have undersized ESP or for installations where ESP performance has been affected by the implementation of other pollution control measures, such as fuel switching. It is important to add that the baghouse systems are designed on either low-pressure/high-volume or medium-pressure/medium-volume pulse-jet cleaning technology, which works online with the system. Due to the fact that the ESP removes the majority of ash or dust prior to entering the fabric filter, the filtration rate (air-to-cloth ratio) can be increased substantially more than the conventional filtration rates, while still maintaining the same pressure drops as that of conventional filtration systems. It is also believed that the ESP serves as a pre-charger and helps to agglomerate the dust particles into a larger and thus far more porous structure over the filter fabric, which also aids in the filtration process. While the true capital cost benefit of COHPAC is achieved at the higher filtration rates, there are several similar installations around the world with fabric filters, which provides benefits in spite of operating at conventional filtration rates [148,237].

In an economic evaluation conducted on two different size units – 350 MW and 550 MW – three PJFF sizes with differing air-to-cloth ratio were selected to cover the full range of possible options. All the pulse-jet filters were installed at the same location, downstream of the existing electrostatic precipitator. In this particular plant, the existing ESPs are installed

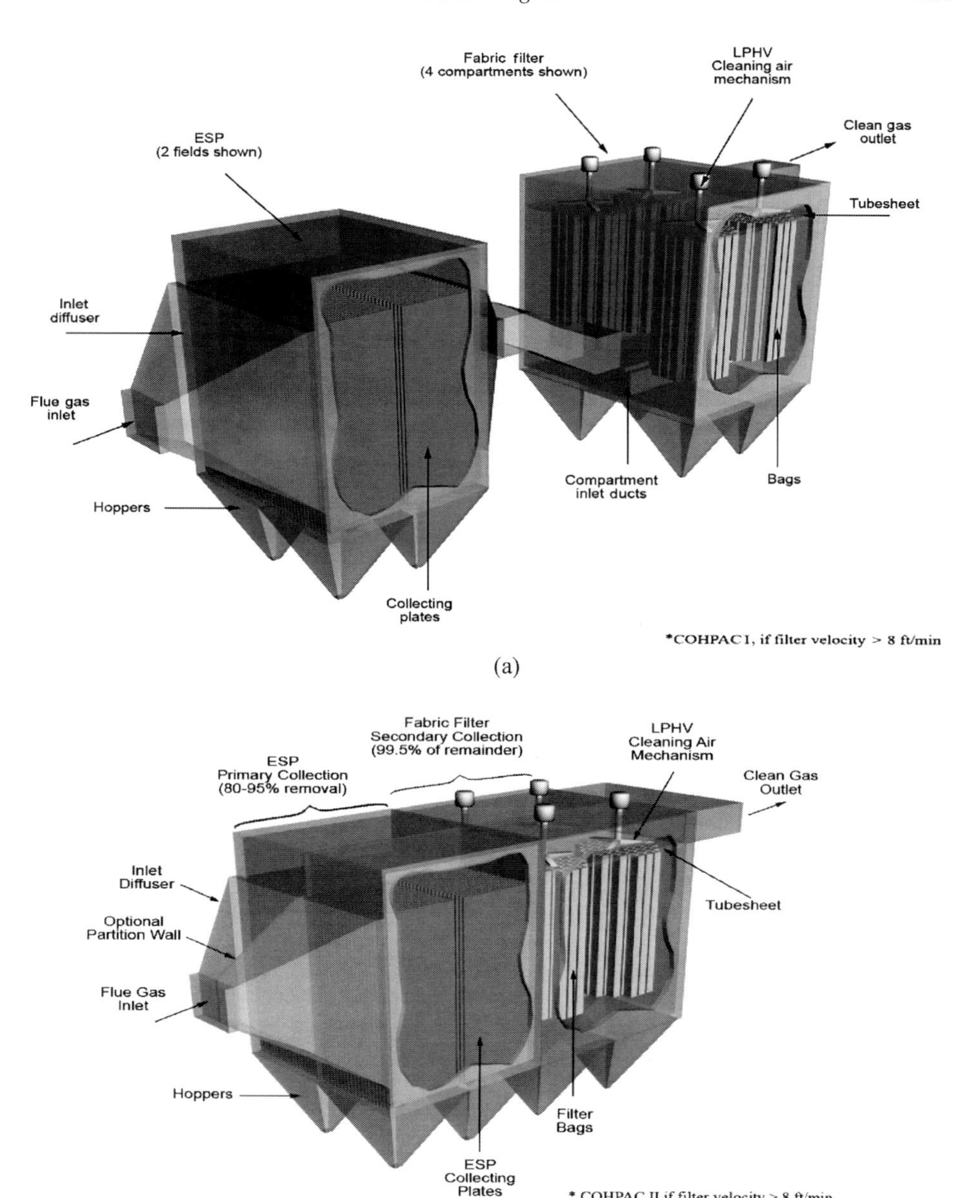

Figure 38. Arrangement for (a) COHPAC I; (b) COHPAC II. (Reprinted by permission of Hamon Research-Cottrell, Inc., USA.)

in a hot-side configuration upstream of the air heaters. With the conventional air-to-cloth ratio, the ESP is deenergized. This results in the air heaters operating in a dirty configuration and necessitates a change to a more open-style basket. The cost of modifying the air heaters is included with the conventional PJFF and is not required with the high-ratio sizes [238].

Conventional system (air-to-cloth ratio: 4 ft/min)

PC boiler → economizer → ESP deenergized → air heater modified → PJFF → upgraded fan → FGD → stack.

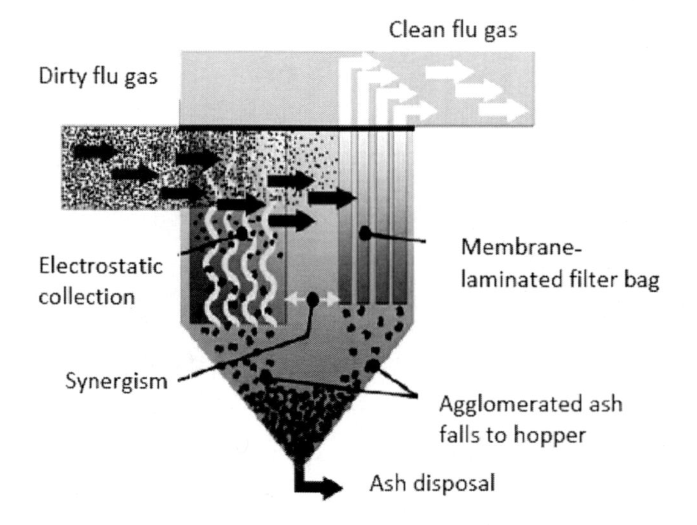

Figure 39. Concept of a hybrid particulate collector (AHPC) [239].

High and ultrahigh ratio system (air-to-cloth ratio: 6 ft/min and 8 ft/min respectively)

$$PC\ boiler \rightarrow economizer \rightarrow ESP \rightarrow air\ heater \rightarrow PJFF \rightarrow upgraded\ fan \rightarrow FGD \rightarrow stack.$$

It was found that when evaluated over a 20-year period, it was clearly more economical to retrofit a conventional pulse-jet fabric filter compared to a high or ultrahigh-velocity filter. In addition, by utilizing a conventional fabric filter additional flexibility is provided, as the operation of the baghouse is not dependent on the continued maintenance and operation of the ESP [238].

The University of North Dakota and the Energy and Environmental Research Center (UND/EERC) have developed a combined particulate control device (Figure 39), which promotes a synergy between fabric filter and electrostatic precipitator. In the filter unit, the filter bags are able to operate at high air-to-cloth ratios and be cleaned without the typical concern with dust re-entrainment. Components such as filter bags made with a microporous polytetrafluoroethylene (ePTFE) membrane, rigid ESP discharge electrodes, and pulse-jet cleaning form an integral part of the overall system, which results in superior particulate control at costs comparable to currently used technologies. Since about 90% of the mass of particles is collected in the electrostatic charging and collection section of the advanced hybrid collector, the load on the fabric filter part of the system is greatly reduced. This filter technology is easily adapted by new installations as well as retrofits of existing ESPs. The design and operation of the filter along with performance data from an operational cement kiln application has been reported [239,240].

A hybrid collector combining electrostatic precipitator and fabric filter was developed for improving particulate matter removal from flue gases from coal power plants followed by a testing in a pilot plant that processes up to 15,000 m^3/hr of flue gas under a simulated situation [84]. The efficiencies obtained fulfils the legal limits in the European Union and the United States of America, with high removal efficiency of PM_{10} (more than 99.95%) and $PM_{2.5}$ (96–98%), and a metal deposition (more than 99%), depending on the metal – overcoming limitations of ESPs with regard to achieving the particulate matter emission limits. However, a lower efficiency was obtained for the capture of Hg in vapor phase (only 30%).

An offline mode appears to be more effective when applied in hybrid collectors. The fly ash load and particle size distribution are related to the pressure drop and the rate of pressure loss in the fabric filter. To achieve the requisite collection efficiencies, the technology requires a total collection area of about 60% of a conventional ESP and about 50% of a conventional fabric. This entails the use of higher velocity along with large decrease in particle load at the inlet of this device [84].

6.2.3. Advanced hybrid particulate collector

In an Advanced Hybrid™ configuration, the internal geometry consists of alternating rows of ESP components (discharge electrodes and collecting plates) and filter bags within the collector [241]. The inlet flue gas is directed into the ESP zone, which removes most of the entrained dust prior to it reaching the filter bags. The perforated collecting plates permit flue gas to pass through them to the filter bags. The bags are placed in a compact arrangement, adjacent to the ESP-collection zone. When the filter bag is pulsed clean, the ESP plays another important role as it effectively captures the dust cake, thereby greatly reducing the potential for dust re-entrainment onto the filter bag (Figure 40). The perforated collecting plate, besides capturing the charged particles, also serves to protect the filter bags from any potential electrical damage from the electric field. The collecting plate and discharge electrodes are periodically cleaned using typical rapping methods [241].

All the flue gases pass through the filter bags, which results in extremely low particulate emission levels due to membrane filtration. The capture of fly ash by the ESP prior to the flue gas reaching the bags and during bag cleaning enables the filter bags to operate at high air-to-cloth ratios that are three to four times greater than conventional pulse-jet fabric filters. Since the Advanced Hybrid™ filter can be operated at high air-to-cloth ratios, less fabric filter components, such as filter bags, filter cages, and pulse valves, are needed. This allows the use of higher performance and more durable components and translates into a more reliable system that requires less overall maintenance. The net result is a cost-effective

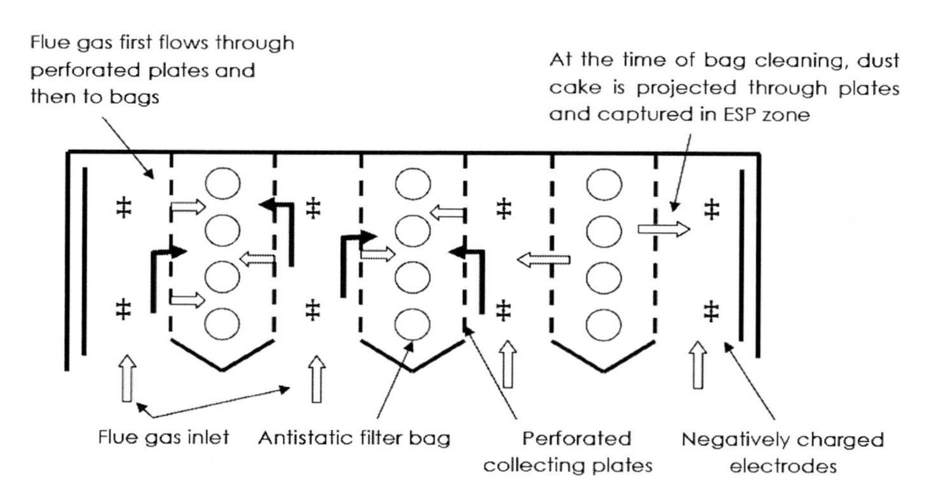

Figure 40. Advanced Hybrid™ filter system [241].

system that offers the reliability of an ESP with the filtration performance of a fabric filter [241].

A series of experiments was carried out to evaluate the effect of several variables, including bag type, electrode type, perforated plate (hole size, opening area, alignment), and operating parameters (current level, pulse trigger pressure drop) on overall advanced hybrid particulate collector performance and bag protection [242]. Results show the perforated plates can provide good protection on three different bags (Gore-Tex® membrane/Gore-Tex® felt filter bag, Gore-Tex® membrane/Gore-Tex® felt, and Gore-Tex® antistatic membrane/GORE-TEX® felt). No electrically induced damage was observed for all three bags based on the visual inspection at the end of the experiments. The perforated plate concept also benefits collector's overall performance by serving as a collection plate to remove fly ash from the flue gas. The perforated plate AHPC system demonstrated long bag-cleaning intervals and low average pressure drop across the system at a low current level (1.5–3.0 mA) and a high air-to-cloth ratio of 3.7 m/min. The residual drag was in the lower range showing an excellent bag cleaning ability under the perforated plate configuration. Results indicated an improvement in ESP collection efficiency with the perforated plate AHPC system [242].

A patent report [243] describes a hybrid filter comprising a plurality of bag filters and plate-shaped grounded electrodes along with high-voltage electrodes arranged between individual filter bag rows such that an electrostatic field is built up on each side of each filter bag row. Due to the high-voltage electrodes and collecting electrodes, most of the particles will accumulate on the collecting electrodes. Only a small portion of impurities will deposit on the outer sides of the filter bags. On account of the filter cake thus growing more slowly on the fabric filter, the dedusting intervals can be extended. Furthermore, during dedusting of bags, particles are transported to the collecting electrode and, for the most part, not attracted again by the outer side of the fabric filter. In order to enhance the bag filter dedusting efficiency, a two-stage dedusting compressed-air pulse is additionally applied, i.e. a short compressed-air pulse of high pressure followed by a prolonged compressed-air pulse of lower pressure. The efficiency of collecting electrodes is enhanced by reversing the direction of the electric field between the electrodes. Further, dedusting of the collecting electrodes can be improved by shaking or beating [243].

6.2.4. Multi-stage collector (MSC)

(a) Compared to other hybrid technologies, the MSC™ provides [244–246] (a) a more compact particulate collector; (b) a more energy-efficient particulate collector even with a high particulate removal efficiency (99.99%) for all particle sizes including fine particles (below 0.1 μm); and (c) the operation of the MSC™ seems to be independent of the electrical resistivity of the collected material, and hence its application would be especially beneficial when electrical resistivity of material either exceeds $10^{11}\,\Omega$ cm or is less than $10^{4}\,\Omega$ cm [244,245].

Figure 41 shows the line diagram of the configuration of a multi-stage collector. The design of MSC™ consists of discharge electrodes placed between oppositely charged collection electrodes. The discharge electrodes are followed by barrier filter elements located in the wide zone placed between the collecting electrodes. Additionally, the surface of the barrier filter element can be made electrically conductive. The corrugated collecting electrodes are formed with narrow and wide sections to accommodate both the discharge electrodes and barrier filter element. They are held at a first electrical potential while the discharge electrodes and the conductive surface of the barrier filter element are held at a

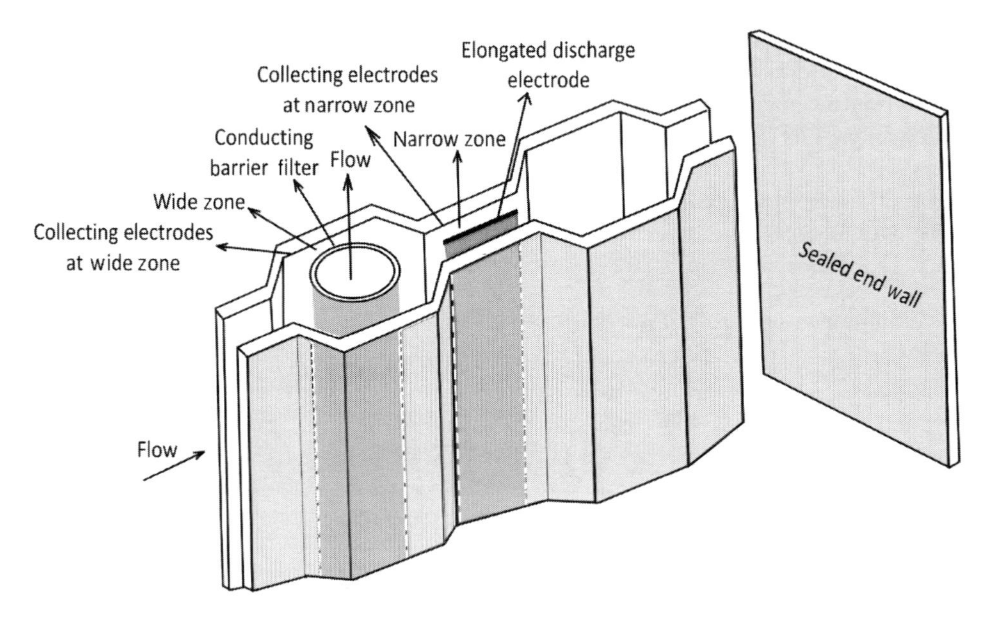

Figure 41. Line diagram of multi-stage collector (MSC™) [246].

second electrical potential. The flat sides of each of the discharge electrodes, collecting electrodes, and the surface of the conductive barrier filter element form collecting surfaces where the electric field is relatively uniform. The sealed end wall prevents further gas flow in longitudinal direction of plates, and forces all gases to flow through the barrier filter. By using an electrode with a cross-section that is relatively wide, a uniform electric field can form in the region of the center of the electrode, and, on the other hand, a nonuniform field of high intensity can form at the sharp curvature of the leading and/or trailing edges. At sufficiently high field strength in the nonuniform field region, a corona discharge can take place between the discharge electrode and the plates acting as an ion-charging source for dust particles passing through it. The center region of uniform field, on the other hand, acts in a manner similar to the field between parallel capacitor plates with charged dust particles collecting on the plates. The dust particles around the discharge electrodes (i.e. in the regions of the corona-generating points), which are charged to negative polarity, are caught by the collecting electrode. Meanwhile, dust particles near the collecting plate electrode, which have been charged to a positive polarity by the positive ions resulting from reverse ionization, are collected by the uniform field-forming part of the discharge electrode and the conducting surfaces of the barrier filter element. Thus, the MSC™ collector can deal with a bipolar distribution of charged particles [244–246].

The system allows for the use of a single high-voltage power source for all electrostatic fields (in all zones). A high-voltage electric field of an adjustable (variable) frequency and/or alternating polarity could also be applied to the device to further improve the collecting efficiency of both positively and negatively charged particles onto the surfaces of the plates, thereby substantially increasing the effective collecting area. All collection surfaces can be cleaned in a conventional manner, such as by rapping, polarity reversal, or by other means. MSC™ is engineered in such a way that the barrier filter element and the discharge electrodes are grounded while the parallel-corrugated plates are suspended from the insulators. Consequently, by virtue of having the barrier filter element at the same potential as the discharge electrodes, the MSC™ design completely eliminates any potential sparks from the discharge electrodes towards the barrier filter element, thus eliminating any chance of causing fires and/or puncturing holes in the porous filter media. Hence, whether the MSC™ is powered by a conventional or an alternating power source, the barrier filter element remains protected from any sparks from the discharge electrodes irrespective of dust concentration [244–246]. However, the above system has not yet been commercialized.

7. Simultaneous controls of particles and gases

Previous technologies associated with industrial gas cleaning have required separate equipment for the respective treatment of dusts and gaseous pollutants (inorganic gases, volatile organic chemicals, etc.) with such systems eventually increasing the cost of industrial products. The recent trend is to control the particles and gases in a singular process step.

7.1. Through sorbent injection

In the process of dry flue gas cleaning, a solid sorbent, mostly $Ca(OH)_2$, is injected into the flue gas in a duct or fluidized bed and the solid is removed downstream by a jet-pulsed filter. A major part of the overall SO_2 and HCl removal of the flue gas takes place in the fixed bed of the filter cake. Some types of processes use the filter alone without a fluidized bed as a reactor. Apart from limiting the emissions of fine particles and gases such as SO_2, large efforts are made to minimize Hg from coal combustion. The type of particulate-control equipment is a key parameter defining the amount of sorbent that is required, and provides the ultimate limitation of the amount of Hg that can be removed. Modern fabric filters are certainly an alternative for an efficient removal of fine particles and offer the option to inject adsorbents to improve the removal of Hg. The concentrations of trace elements like As, Cd, Co, Cu, Pb, and Sb in different particle sizes can also be minimized. It is important to note that the co-combustion of waste-derived fuels is leading to stricter limitations regarding particulate emissions. This is due to incineration products like furan and dioxins that are known to adsorb on solid substances.

The combined use of fabric filters with sorbent injection systems has been utilized for many years in the municipal incinerator as well as in other industries as a way to enhance the removal of Hg and other pollutants such as dioxins, furans, and a wide range of heavy metals. The dust cake, which forms on the filter from the collected particulates, can significantly increase particulate- and Hg-collection efficiency. With a carbon injection system upstream of the fabric filter, the carbon-enriched dust cake on the fabric serves as a fixed bed reactor, providing excellent contact between the Hg-laden flue gas and the reactive carbon. A fabric filter provides a relatively long residence time of several minutes compared to an ESP, which may only have 2–3 seconds of *in-flight* exposure. The difference in gas phase diffusional contact between *in-flight* sorption and fabric filter cake contact can be evaluated. In the cake, the particle–particle distance is much shorter, of the order of 30 μm, and the ratio of the diffusion distance to the particle–particle distance is much higher for the filter cake. Therefore, the diffusional mass transfer from the gas-phase to the particle is more efficient in a fabric filter cake than during the *in-flight* period dominating the ESP case. This explains why there is a factor of 10 differences in carbon feed rates required for ESPs as compared to fabric filters [148,237,247].

Coal-fired power plants are the most difficult Hg-emissions sources facing control under the Clean Air Mercury Rule (CAMR) issued by the U.S. EPA. Unlike municipal and medical waste incinerators (MWIs), which emit much higher concentrations of Hg that is almost exclusively in the oxidized form Hg^{2+}, CFPPs emit very dilute (single ppb) concentrations of both Hg^0 and Hg^{2+}, whose proportions in relation to the total Hg load can vary widely. Since Hg^{2+} is more condensable and far more water-soluble than Hg^0, there is large variability in Hg in the total Hg removal efficiency of most Hg-emission control technologies. Although the most mature control technology is adsorption across a dust cake of powdered sorbent in a fabric filter, most particulate control in the United States associated with coal combustion takes the form of ESP [55].

Activated carbon is a representative sorbent for Hg removal in flue gas. In Hg adsorption by activated carbon, physical and chemical adsorption mechanisms are all known to be effective. The use of activated carbon having large surface area can enhance the physical adsorption of Hg. Further, impregnation of activated carbon with elements such as sulfur, iodine, chlorine, and bromine can promote the chemical adsorption of Hg [248–251]. These elements are known as providers of active sites for Hg bonding on a carbon surface. Presently, among various forms of activated carbon (e.g. brominated carbon, sulfur-impregnated carbon, and chloride-impregnated carbon) and carbon substitutes (e.g. calcium sorbents, petroleum coke, zeolites, and fly ash), brominated activated carbon appears to be the best-performing Hg sorbent. It is important to note that there is also a possibility of a noninjection regenerable sorbent technology [252].

Effects of activated carbon injection rate on Hg^0 removal in a bench-scale particulate collector with fabric filters were experimentally estimated, and the contribution of activated carbon collected on a filter surface to Hg^0 removal was evaluated [251]. At given conditions, Hg^0 removals converged to a certain level as activated carbon continued to be injected irrelative to the C/Hg ratio or the type of activated carbon. It was reasoned that possibly the particulate collector acted not only as an additional adsorption reactor but also as buffer storage to accumulate activated carbon particles inside to a certain amount [251].

Furthermore, the contribution of activated carbon collected on the filter surface to the removal of Hg^0 was not very large. This may be ascribed to the short contact time for interaction between Hg^0 and the activated carbon collected on the filter surface compared to that distributed inside the chamber. From the results of the activated carbon injection experiments, it was interpreted that the Hg^0 concentration in a particulate collector was reduced mainly by the activated carbon distributed inside the chamber rather than that collected on the filter surface [251]. This enhanced filter cake provides higher inherent removal of Hg with much lower required sorbent injection rates, thus reducing the overall operation and maintenance costs associated with sorbent injection for Hg control. However, the result could change to some extent when the conditions applied, such as flue gas composition, filter-cleaning timing, temperature, and the inlet concentration of Hg, are changed. Particularly in the case of the recent particulate collectors having a high-filtration velocity option in which multiple mechanisms for particulate collection are merged, the effect of activated carbon collected on the filter surface on the removal of Hg^0can be smaller than that of conventional baghouse [251].

For the removal of Hg and particulate matter, fabric filter technologies can be located in a multitude of locations, including being used as the primary particulate control device or downstream of an existing or new ESP collector. Some designs place the fabric filter within the ESP casing in the design of a full ESP conversion to a pulse-jet fabric filter or even a partial conversion with the last few fields being converted to a high ratio pulse-jet. The type of coal has been clearly shown to aid or diminish the ability to remove Hg with activated carbon injection control. Although fuel additives can be beneficial in enhancing the ability to control Hg, but such additives may be detrimental to the boiler, air heater, or downstream

Figure 42. TOXECON set-up combining ESP, activated carbon injection (ACI) and pulse-jet fabric filter (PJFF) (*developed by Electric Power Research Institute, USA and ADA - Environmental Solutions, USA*).

equipment. Since carbon does have a negative influence on cake resistance, with higher proportion of active carbon injection rate, the working air-to-cloth ratio will be reduced [238,247]. Hence, carbon injection rate is required to be monitored all over the process.

One of the most popular technologies in Hg control is TOXECON (Figure 42). The technology is strictly within the Electric Power Research Institute (EPRI) patents, which would require either working with ADA Environmental Solutions (who are generally in the Hg technology business) or getting a separate license from EPRI. The development of TOXECON technologies was based on two needs. The first being able to enhance poorly operated ESP by adding a polishing type fabric filter downstream of the primary EPS collector, and second, by adding activated carbon inbetween the two collectors, it is possible to reutilize the fly ash collected in the ESP as an additive for concrete production and sale to cement producers. When activated carbon is added to fly ash, the higher loss on ignition adds a darker color to the product, which in itself may reduce its attractiveness for resale, plus the activated carbon tends to reduce the oxygen levels in the concrete, which causes it to fail the foam index tests and can reduce the concrete's overall strength levels.

There are several forms of TEXCON configuration; it can be even with mechanical system. But the filtration rate would likely be reduced as the mechanical collector is not very efficient in the removal of particulates and, as a result, the majority of particulates would be captured downstream in the polishing (TOXECON)-type collector. Usually TOXECON can be considered a sorbent injection analogue technology to COHPAC. TOXECON I is an addition to a secondary collector (mostly a fabric filter system) downstream of the primary ESP collector. In some applications, like in cement kilns, this could involve adding a fabric filter in series with the primary fabric filter to be used only for toxins' collection and capture on activated carbon. This is used for the majority of polishing-type collector applications.

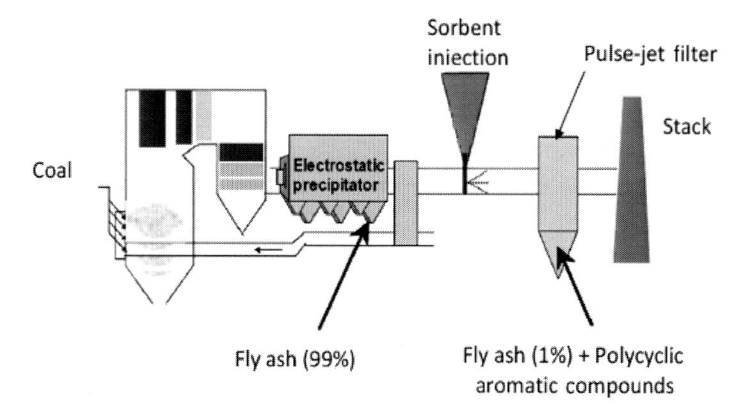

Figure 43. Schematic representation of TOXECON configuration [237]. (Reprinted by permission of ADA – Environmental Solution, USA and Electric Power Research Institute, USA.)

In one of the preferred configurations, the use of dry additives, such as powdered activated carbon (PAC) in combination with COHPAC, also known as TOXECON (Figure 43), has been demonstrated to achieve high Hg-reduction levels on both coal-fired and waste-to-energy combustors. The system represents one of the most cost-effective approaches of reducing Hg emissions from coal-fired boilers at relatively low carbon feed rates, without contaminating the bulk of the ash. It may be added that the filter velocity range of 8 to 40 ft/min for COHPAC is not a comfortable range for many utilities, and application of a polishing filter (whether in the single casing or separate) at less than 8 ft/min is not considered for COHPAC; with ACI injection for Hg in such a polishing filter is not TOXECON by the definition of compact baghouse. Injection of ACI between the fields of an ESP is considered TOXECON II [253], and this has been one of the development targets of ADA in field-testing at Entergy Independence Station. TOXECON II^{TM}technology requires minimal capital investment because it uses only minor retrofits to the ESP for the carbon injection system instead of installing a separate secondary particulate control device. However, the system is still under development stage.

Assuming that the fly ash is currently being sold by the power station, even a small amount of carbon can significantly reduce or eliminate the ability to sell the collected fly ash. With either COHPAC I or COHPAC II configurations, the ash streams can be separated and the smaller portion of the ash being captured in the fabric filter system can be sent to waste disposal while the ESP ash can continue to be sold. The use of EPRI's TOXECON technology in association with COHPAC can provide additional benefits, including reduction of both total and fine particulates, Hg reduction via use of activated carbon, and potential reductions in SO_2 and HCl levels with the addition of alkali agents such as sodium or hydrated lime [25,237].

7.2. Catalytic filter

The growing need for energy and space savings has forced chemical engineers to work out new *multi-functional reactors* capable of carrying out, besides the chemical reaction, other functions such as separation, heat exchange, momentum transfer, secondary reaction, etc. [254,255]. A common feature of all multi-functional reactors is that they allow substitution of at least two process units with a single one, where all the operations of interest are carried out simultaneously. A likely consequence is the reduction of investment costs, which is often combined with significant energy recovery and/or saving. In the development process, catalytic filters are found to be capable of removing particulates from flue gases (e.g. from waste incinerators, pressurized fluidized bed coal combustors, diesel engines, boilers, biomass gasifiers, etc.) and simultaneously abating chemical pollutants (e.g. nitrogen oxides, dioxins, VOCs, tar and carbonaceous material, etc.) by catalytic reaction. The catalyst is applied in the form of a thin layer directly onto the constituting material of the filter, which can be either rigid (filter tubes made of sintered granules) or flexible (ceramic or metallic fiber tissues) [256,257]. However, the following properties should be exhibited by catalytic filters so that these opportunities can actually be exploited:

- Thermochemical and mechanical stability.
- High–dust-separation efficiency: dust should not markedly penetrate the filter structure since this would lead to pore obstruction and/or to catalyst deactivation.
- High-catalytic activity at conveniently high-superficial velocities employed industrially for dust filtration (10–80 $Nm^3\ m^{-2}\ hr^{-1}$).
- Low pressure drop even with the presence of a catalyst.
- Low cost.

In Japan, a technology for simultaneous removal of dusts and gases has been studied recently using a catalyst bag filter, and work on the advanced oxidation process has been conducted to treat volatile organic chemicals and odors in Korea [258]. Through combining the principle of surface filtration and catalysis [259], the new methodology destroys gaseous dioxins and captures solid phase emissions in applications like incinerators, pyrometallurgical, and cement kilns. All of these benefits translate into operational cost savings.

The combination of particle filtration and catalytic filter in a single unit is based on certain fundamental requirements. In the industrial filtration process, efficient removal of nitrogen oxides can be achieved by using the selective catalytic reduction (SCR) technique. SCR is a nitrogen oxide-reduction process where NOx in the flue gas is selectively reduced to nitrogen and water in the presence of oxygen using ammonia as a reducing agent. The reduction occurs in the presence of a catalyst at a reaction temperature typically between 250°C and 350°C. In most cases, honeycomb structures are used as catalyst unit in SCR systems. In some special applications, plates or fixed beds are also used. The SCR system can be located upstream or downstream from the particulate control device. In case of a high dust SCR system, hot dust-loaded gas passes through the SCR catalyst unit. In case of a low-dust SCR system, particulates are separated before passing through the SCR catalyst unit. In combustion plants, an additional desulfurization step is usually performed before entering the SCR system depending on the kind of feedstock used [260,261]. The main disadvantage of a low-dust SCR system is re-heating the particle-free gas to the required SCR catalyst-operating temperature. On the other hand, the main disadvantage of a high dust SCR system is that catalyst plugging takes place.

In order to overcome these disadvantages, the combination of filtration and SCR reaction in one unit using a catalytic filter [259–265] allowed the use of the high-energy content of the gas as well as prevents plugging of the catalyst. Furthermore, the combination of two units into a single unit reduces processing as well as investment and maintenance costs. Fabric filter or ceramic hot gas filter elements (foam candle) with a fine filtering outer membrane and a catalyst integrated in the support structure of the filter elements can be used to achieve an efficient particle as well as an efficient NOx removal (Figure 44). The use

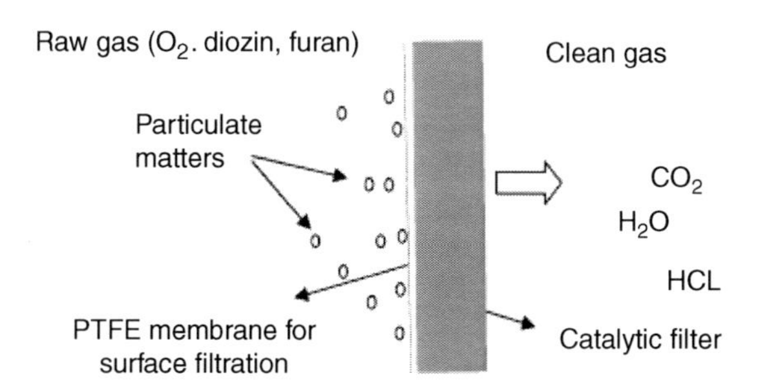

Figure 44. Combining the principles of surface filtration and catalysis.

of these filter elements enables us to realize the combination of a filter and an SCR reactor in a single unit. Moreover, the function of the integrated catalytic layer can be tailored in such a way to allow not only catalytic NO_x removal but also the catalytic oxidation of VOC [259,266,267]. Since PTFE lamination can withstand a continuous temperature of 260°C, it is useful to develop a catalyst active in such a temperature for the SCR of NO_x with ammonia and simultaneous combustion of VOCs [259].

Earlier Spivey [268] reported on the use of catalytic baghouse filters used in Japan to control the pollutant levels in flue gas from municipal waste incinerators. In one of the systems, the medium is made by impregnating the glass-fiber fabric with a vanadium pentoxide catalyst. A 25% solution of ammonia, slaked lime, and a fire clay powder (whose role is to form a protective film that prevents the bag, fabric from clogging) are injected into the flue gas after it has passed through a quenching chamber. During filtration, SO_2 and HCl react with lime to form a cake, which is removed periodically from the surface of the filter bag while NO_x reacts with NH_3 on the filter itself. NO_x levels are reduced from 109–150 ppm to 30 ppm, and HCl, SO_x, and fly ash levels are reduced by 98%, 88% and 99%, respectively. Dioxins and Hg are also removed by the reaction and absorption process respectively.

In order to comply with dioxin and furan (PCDD/F) emission regulations in Belgium, the IVRO municipal waste incinerator adopted catalytic filters for use in the plant's two existing fabric filters [269]. This system replaced the injection of PAC because of concerns that PAC, used at temperatures above 200°C, would ignite and lead to fires and plant downtime. The performance of the catalytic filter system, since its installation in 1997, has been reported. The system shows that greater than 99.5% of the gaseous PCDD/F entering the fabric filter is destroyed by the catalyst within the filter media, and PCDD/F emissions are found to be well below the regulatory limit of 0.1 ng I-TEQ/Nm^3.

Based on different catalytic actions, nonwoven treated with photocatalyst was found to be effective for simultaneous removal of particles and gases [358]. Photocatalytic oxidation is considered to be one of the most effective ways to decompose various chlorinated alkenes and other volatile organic chemicals of low concentrations in gaseous streams. Especially, photocatalyst oxidation belonging to the advanced oxidation process is a heterogeneous reaction producing an OH radical that decomposes organic pollutants to form nonhazardous final products such as carbon dioxide and water. The OH radical reacting with the most organic materials has an oxidation potential superior to that of other existing oxidants, and its reaction rate is very fast [270,271]. As a supporter for the photocatalyst, nonwoven fabric is capable of attaining sufficient active sites, homogeneity, and the titanium dioxide (TiO_2) catalyst adhesion to the surface [272]. The filtering materials of nonwoven fabric type have wider surface area than those of film or pack type. Thus, the coating treatment of nonwoven fabric with TiO_2 maximizes the functional effectiveness of filtration and photooxidation.

The behavior of nonwoven fabric coated with TiO_2 photocatalyst for the removal of dust particles and VOC from a contaminated air steam was reported [258]. A fabric filter sampling system was manufactured by using a UV lamp and a plate-shape sample fabric coated with TiO_2 sol to develop a photoreacting fabric filter. Variations of pressure drop across the fabric as well as toluene vapor removal efficiency were investigated and examined with respect to various conditions such as the injection duration of dust-laden gas, the photocatalyst particle size, the toluene vapor load, and the photocatalyst load. Further, variations in air permeation and the tensile strength of fabrics with photocatalyst load were measured and examined to determine the appropriation of nonwoven fabric as a supporter for photocatalyst. The toluene removal efficiency of this new generation fabric filter was manifestly significant, and a possible deficiency in removal efficiency would be solved through appropriate design and serial arrangement of a multi-channel photoreactor consisting of fabric media coated with a photocatalyst [258].

An innovative catalytic filter concept based on the combination of a laminated fabric filter for fly ash filtration and catalytic foams for the reduction of NO_x and combustion of polycyclic aromatic hydrocarbons present in incinerator flue gases, was conceived [257]. Figure 45 shows the concept of a multi-functional filter. The feasibility of the concept was tested by developing suitable catalysts and depositing them inside ZTA ceramic foams. Catalyst development with combinations of MnO_x, · CeO_2, and V_2O_5 WO_3, and TiO_2 catalyst species were found to be successful for reaching high NO_x and VOC removal efficiencies (>80%) at the filter bag operating temperatures (200–210°C) and for superficial velocities of industrial interest (10–60 Nm^3 m^{-2} hr^{-1}). Catalyst deposition inside the selected foam substrates was operated by impregnation with a catalyst suspension, followed by microwave drying and calcinations. Some detrimental effect of SO_2 on the catalyst activity suggests its removal at the initial stage either by dry scrubbing or other suitable technique. For effective reactor performance, even catalyst distribution within the foam is essential [257].

Heidenreich et al. [273] reported the one-step dry flue gas-cleaning process using rigid ceramic catalytically active filter elements for the combined removal of particles and nitrogen oxides as well as the development of these filter elements. The process is based on a multi-functional filter, which combines particle filtration and selective catalytic removal (DeNOx reaction) by using rigid ceramic catalytically active filter elements. In the study, catalytic filter candles are made from a coarse porous support body based on silicon carbide grains with a fine filtering membrane of mullite grains (Figure 46). The mean pore size of the support body is 50 μm. At the upstream, different membranes with different pore sizes can be provided depending on the field of application and demanded filtration efficiency. The fine membranes achieve filtration of particles with sizes down to less than 0.3 μm. Additionally, the filtering membrane protects the porous catalytically coated filter support body against deactivation by particle deposition, which is very important for the life of a catalyst.

By injection of sorbents upstream of the catalytic filter, gaseous pollutants and potential catalyst poisons such as SO_x and HCl can be removed. The pollutants SO_2 and HCl are removed by using sodium bicarbonate ($NaHCO_3$) or $Ca(OH)_2$ as sorbent, whereas in the presence of NH_3 and O_2, NO_x is catalytically converted to N_2 and H_2O by passing through the catalytic filter elements. Due to the high residence time of the sorbents at the filter surface, gaseous pollutants (SO_2 and HCl) are efficiently removed before entering the catalytic layer. In this way, a completely dry flue gas-cleaning system (Figure 47) can be realized. By periodically back pulsing, the dust (fly ash) and deposited sorbent are detached

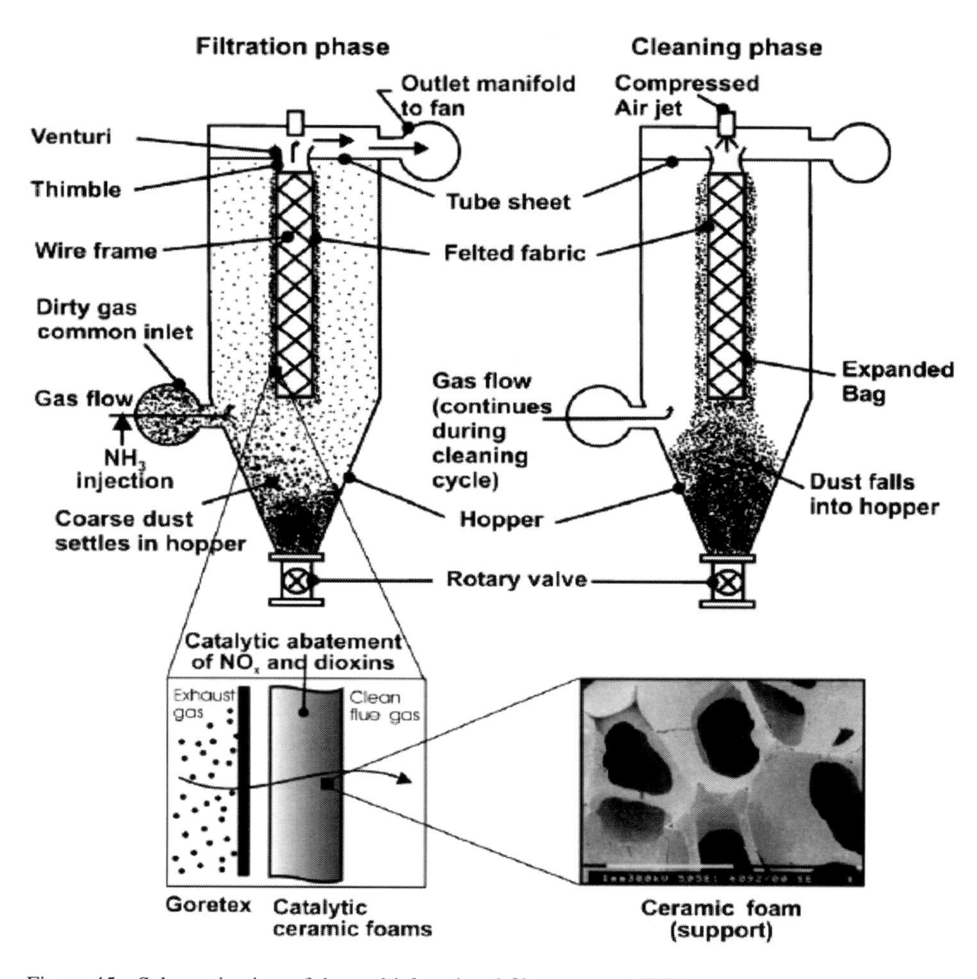

Figure 45. Schematic view of the multi-functional filter concept [257].

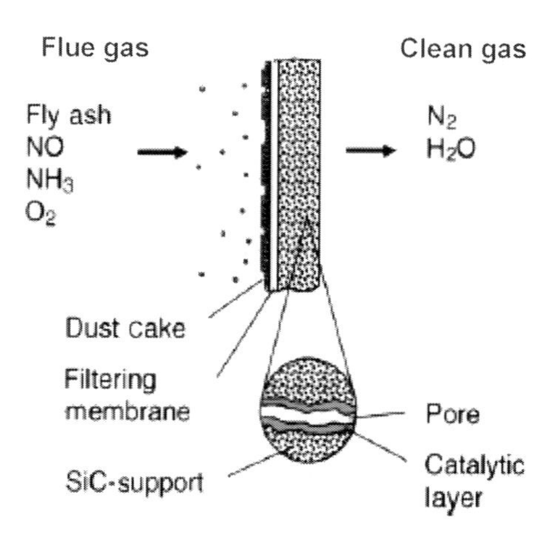

Figure 46. Structure of the catalytic filter element [273].

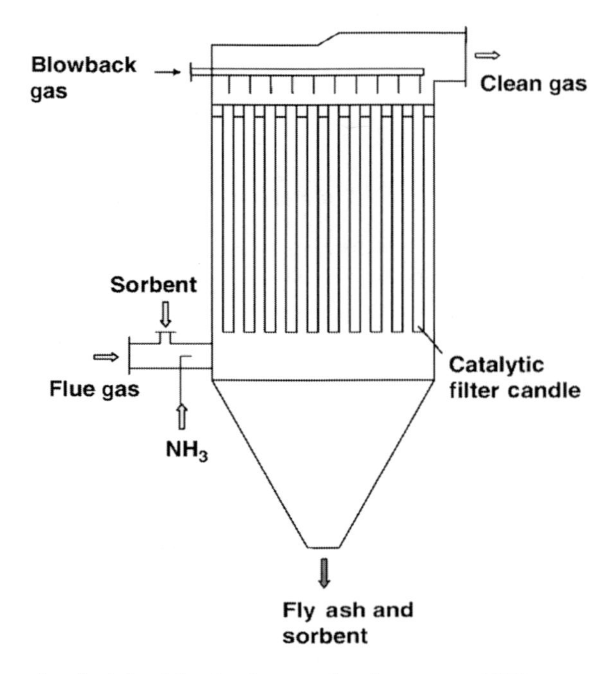

Figure 47. Schematic principle of the dry flue gas cleaning system [273].

from the filter element surface, collected at the bottom of the filter cone and periodically discharged from the system [273].

The above catalytic filter elements were tested under different operating conditions, such as filtration velocity, operating temperature, and nitrogen oxide inlet concentrations. Moreover, the filtration properties of the filter elements were determined. Through these studies, the optimum operating conditions for the catalytic filter element are determined with respect to a minimum differential pressure value in combination with a high-catalytic activity towards NOx removal [273].

7.3. Using nonthermal plasma

Technology based on nonthermal plasma (NTP) is expected as a new energy source to activate chemical reaction in gas phase, especially in gaseous pollutant treatment reaction. Nonthermal plasma is characterized by the fact that the electron temperature is very high whilst the ion and neutral molecular, i.e. the gas temperature, is low. Nonthermal plasma, therefore, has merits of applicability to the materials and conditions that are intolerant for high temperatures. Since the equipment can run under moderate conditions like normal pressure and room temperature, the apparatus configuration is simple and no vacuum system is required. Nevertheless, it can activate chemical reactions that must not progress unless at a high temperature even under normal temperature and pressure. An NTP in flue gas is a quasineutral mixture of charged particles (electrons, positive, and negative ions) and chemically active particles like radicals and photons. Photons are created from the collisions of energetic electrons with molecules of background gas. A very useful property of an NTP is that the majority of the electrical energy is utilized in heating the electrons rather than heating the gas. In the case of a gas flow containing N_2, O_2, and H_2O, most of the primary radicals generated in an NTP are O and OH. These radicals rapidly oxidize NO and SO_2

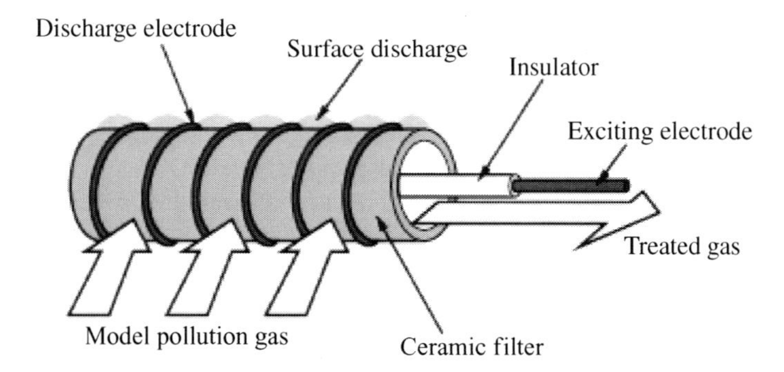

Figure 48. Schematic illustration of the electrode system used in the experiment [275].

to form NO_2 and SO_3, which become nitric and sulfuric acids through reaction with H_2O. These gaseous acids can be transformed into salt with gaseous ammonia. Subsequently the salts are removed as solids. Furthermore, joint gas phase reaction of radicals and ammonia with gaseous Hg can effectively transform 80–90% of Hg into fine particles that can be collected by precipitator or fabric filter [274–276].

Two different nonthermal plasma technologies, such as pulse corona-induced plasma chemical process or surface discharge-induced plasma chemical process [277,278], can be used for the effective removal of gaseous pollutants. Harada et al. [275,276] propose a newly developed electrode system for gas treatment, in which a surface discharge-induced plasma chemical process and ceramic filter are integrated. This reactor system can be characterized by a very simple configuration, in which a thin wire is wound up around the surface of a cylindrical ceramic filter. In the system, the gas to be treated is forced to pass through the ceramic filter on which the surface discharge is applied (Figure 48). Accordingly, very high contact efficiency between the gas and the surface discharge plasma is accomplished compared to the conventional surface discharge reactors. In addition to the high contact efficiency, the ceramic filter can capture particulate pollutants, which will deposit on the surface of the ceramic filter. Because the surface discharge-induced plasma chemical process is applied to the deposited solid pollutants on the surface, gaseous and particulate pollutants suspended in the gas can be treated simultaneously in the reactor system.

The feasibility of the newly developed apparatus was studied for NO_x and soot treatment [275]. NO_x as a gaseous pollutant was decomposed effectively. Soot suspended in gas as a particulate pollutant was captured on the ceramic filter and disappeared with the surface corona discharge treatment on the filter surface. Because soot is removed by oxidization and NO_x by reduction, there is a possibility that they can be treated simultaneously and complementarily. Further, the feasibility of the reactor to decompose VOCs and performance were evaluated [276]. The efficiency was examined against various conditions of initial concentration of VOCs, as trichloroethylene, toluene and benzene, and using variable flow rates of carrier gases (air, nitrogen, and mixture of nitrogen and oxygen or water vapor). Trichloroethylene was decomposed effectively within the reactor system, especially when the carrier gas contains oxygen or water as an oxygen source, whilst the decomposition efficiency was low with the nitrogen carrier. With a sufficient oxygen source, no organic by-products were generated in the treated gas, and trichloroethylene was oxidized to inorganic substances. Toluene and benzene were efficiently decomposed as well [276].

Multi-stage collectors (Figure 41), originally developed for particulate collection [246], can be modified for simultaneous control of gases and particulate pollutants. In the system [274], pollutants are passed through an alternating series of glow discharge regions, constant field collection regions, and corona streamer discharge regions in narrow and wide areas defined by spaced electrodes. It allows the use of a single DC or AC power supply to provide different types of discharge along with linear electrical field collection. The system causes steady state dissipation in said regions of nonpulsed electrical power of sufficiently high voltage to establish therein a continuous glow and a streamer corona discharge. There is a uniform electrical field in wide sections for collecting particles charged by the glow discharge (placed in some narrow section); consequently, particles are removed by the collector electrode and barrier fabrics. On the other hand, streamer discharge electrodes in wide or narrow sections convert and destroy harmful gaseous pollutants. Steady state streamer discharge can convert the polluting gases (SO_2 and NO_x) into nitric and sulfuric acid, which can be removed by combining it with ammonia to form solid salts. Furthermore, joint gas phase reaction of radicals and ammonia with gaseous Hg can effectively transform 80–90% of Hg into fine particles that can be collected by precipitators or fabric filters. It also decomposes volatile organic compounds during the process. It may be added that NTP system in the present configuration does not produce much ozone. Further research in the near future should involve the problems of operating conditions, power efficiency, and capacity of the treatment so that the technology could be employed in practical filed operation.

8. Conclusions

The industrial air pollution control emerges into a very effective gas-cleaning method, which belongs in the basic requirements for atmospheric environmental activities, as well as in the processes of production of new high-tech materials and machine design. The new results of aerosol filtration science will have an important impact on further environmental and technological developments. Through environmental control, not only particulate matters are separated but also various harmful gaseous matters are suppressed. A survey of concentration limits for emissions of and for exposure to a number of pollutants is given in Table 12 [15]. The table shows that there are very large differences between the ratios of emission and exposure limit concentrations for individual pollutants. For example, the limit concentration for particulate matter is set

Table 12. Emission and exposure limits for various pollutants [15]. (Reprinted from D.V. Velzen, H. Lagenkamp, and G. Herb, *Review: Mercury in waste incineration, waste management and research*, 20 (2002), pp. 556–568, with permission of Sage Publication Ltd., UK.)

Pollutants	Emission limit (EM) mg/Nm^3	Exposure limit (EX)	Ratio (EM/EX)
Particle matter	10 mg/Nm^3	125 μm/Nm^3 (24 hr)	80
SO_2	50 mg/Nm^3	125 μm/Nm^3 (24 hr)	400
HCl	10 mg/Nm^3	8 mg/Nm^3 (5 ppm)	1.25
CO	50 mg/Nm^3	10 mg/Nm^3 (8 hr)	5
Hg	50 mg/Nm^3	50 mg/Nm^3 (8 hr)	1

at 10 mg Nm^{-3}, whereas the maximum admissible concentration (MAC) at work places is 125 μg Nm^{-3}. The ratio between both values is 80. On the other end of the scale is Hg, where the concentration limits for exposure and emission are identical, i.e. 50 μg Nm^{-3}. The reason for the large difference in the emission limits between Hg and the other toxic compounds is not clear. The above ratio will be more when the dispersion phenomena occurring during the emission of a flue gas into the atmosphere are taken into account. The flue gas from waste incinerators undergoes very rapid dispersion and dilution after leaving the incinerator stack. It follows that the maximum Hg concentration in the ambient air will remain at least five to six orders below the magnitude of the lowest MAC value and public health will not be threatened. On the other hand, the environmental impact of the sophisticated treatment of flue gases for Hg control comprises an increased amount of hazardous waste, the operation of more complicated treatment processes, and the risks of handling an inflammable material such as active carbon. Future studies must address the above issue very carefully considering pollutants like SO_2, NO_x, furan, dioxin, etc., other than Hg alone [15]. It is also very important to develop technology for the prevention of emission, the act relating to closed substance cyclic waste management and environmentally compatible disposal of waste covering the areas of avoidance, recycling, and waste disposal.

Nevertheless, with the increasingly stringent emission regulations of fine particulates and air toxics, pulse-jet fabric filters have become an attractive particulate-collection option for utilities. PJFFs can meet stringent particulate emission limits (over 99.99%) regardless of variation in operating conditions [279]. The increasing popularity of the pulse-jet system is also due to online cleaning applications, often without compartmentalization, outside the collection which allows the bag maintenance in a clean and safe environment. However, more stringent air pollution laws, especially when using waste derived from fuels, are drawing more attention to the off-gas cleaning facilities, usually bag filters. Pulse-jet fabric filtration systems can cope with most industrial gas cleaning problems economically and efficiently. Filtering of hot gases before heat exchange, catalytic oxidation in combination with bag filters, electrostatic precipitator in fabric filter with conductive filter media, dry scrubbing to blind the gaseous components, filtration with very fine difficult particulates, are just some of the uses. A relatively new technology, but one which is being increasingly applied in high-temperature air pollution control, is the use of low-density ceramic filters in filter plant [280].

With respect to the overall design of a filter unit, there exists a lot of variation in the construction depending on manufacturer and application needs. For separation of difficult particulates, changes in basic design of filter unit and also the operating parameters are quite common. Although some advanced filters, such as hybrid technology, have come into place for specific applications, more commonly existing ESP may be completely converted into a fabric filter at conventional filter velocities (0.9 to 1.4 m/min) as an 'ESP/FF conversion', or a brand new 'standalone fabric filter' is built. This decision is typically made based on site or schedule constraints. Furthermore, choosing different classes of pulse-jet filter (LPHV/MPML/HPLV) is a matter of application needs, operating and maintenance cost, easy installation, etc. In pulse-jet filters, one of the basic requirements is to minimize the operation cost. The energy used by the fan accounts for 60–80% of the baghouse operation costs [171] and, therefore, a stable and low differential pressure makes it worth investing in highly developed filter unit. This involves improved system design, judicious selection of filter media, and setting of operating parameters at the optimum level.

During pulse-jet filtration, regulation of pressure drop is accomplished through optimizing the impulse used and extent of cleaning of filter media in each cycle. Cleaning should not damage the bag filter while allowing filtration processes to operate at a steady and lowest possible pressure drop. It is also necessary to conserve the dust layer up to a certain extent to ensure good filtration efficiency and, in certain cases, helps in absorbing gas on dust cake of specific properties. It may be noted that filter medium has to be designed/selected based on the application. The various filter media proposed for same application can have different flow rate characteristics and air-cleaning efficiencies. The most effective way to determine the correct media for the application and dust is through testing. Acquiring emission and pressure data along with particle size analysis before and after filtration while filter medium is run through a small-scale simulation is useful to determine the effectiveness of the filter. It is also important to note that the filtration requirements vary depending on the nature of aerosol. The media efficiency required for carbon black will probably not be needed for wood shavings [99].

Apart from satisfying stringent emission requirements at varied situations, filter media should have many other properties such as resistance to temperature and chemical environment, dimensional stability, resistance against flexing fatigue and abrasion, etc. depending on the type of application. The filter media should be deigned to minimize the inevitable changes in pressure loss and face velocity during cyclic operation of filtration and dedusting. Designing of filter media is therefore primarily aimed at enhancing its life while maintaining the emission requirement. However, in most of the cases, life of the filter is limited by trapping of pores by small dust particles influenced by the structure of the filter media, filtration velocity, and/or damage caused by pulse cleaning. Deciding air-to-cloth ratio is one of the most critical tasks for overall system design and successful running of filtration operation.

In the development of industrial filter media, a common approach is enhancing surface filtration. Although basic nonwoven structure has been improved significantly, further enhancement of surface filtration is often accomplished by various means like using membrane or coating; or employing fine fibers, trilobal/multilobal fibers at the upstream side of layered nonwoven fabric, etc. Out of the several techniques, membrane filters have become more common for controlling fine particulates. Notwithstanding the use of PTFE, filter membrane increases the cost of the product. There is also a need for looking at their mechanical strength and lifespan. Even small membrane damage will allow dust penetration into the fabric, followed by further delamination and will cause increase in the pressure differential. Higher dust loads, uneven gas distribution, excessive cleaning pressure, incorrectly sized cages, or, in some cases, carelessness during installation will lead to membrane defects and consequently performance failures. One of the major challenges is the development of strong membrane without affecting permeability characteristics. The membrane represents probably the fastest growing part of the filtration media market. The future will undoubtedly see more advanced products of this type, leading to structures that are efficient in particle capture, provide longer filter life, and also capable of performing under more chemically and thermally challenging environments. All of these benefits translate into operational cost savings and better environmental pollution control.

Regarding the overall system design, there are some unique developments which have taken place over the past decades. There are several hybrid filters based on the combination of cyclone/electrostatic mechanism fitted with a pulse-jet bag filtration system. Advanced hybrid particulate collector combines the best features of ESPs and baghouse, providing major synergism between the two collection methods, both in the particulate collection step and in the transfer of dust to the hopper. There are also technological developments for simultaneous control of gaseous matter. Existing technology for gas cleaning at high temperature involves candle filters for removing solid contaminants and sorbents for removing fluid contaminants. Of all future combustion processes, out of the available cleaning methods, rigid ceramic filters have emerged as a promising technology for the cleaning of hot gases due to their resistance to attack by aggressive gases and high temperatures (e.g. in pressurized fluidized-bed combustion and gasification at temperatures up to 1000°C). The COHPAC system also shows promise for location on the hot-side of EPSs, by placing ceramic filter elements in the last ESP field. COHPAC has the potential to upgrade utility's pollution control measures, most likely including SO_2 removal as well as reducing Hg, and possibly NO_x emissions. However, the whole process should be cost-effective and must compete with all other options currently being evaluated. COHPAC could well offer a very interesting and cost-effective solution to future problems in coal-fired utilities. It appears that by 2015 coal-fired electric utilities will dominate the industry, followed by nuclear, hydro-powered, natural gas, and oil-fired stations. Thus, fabric filter technology based on the COHPAC system carries a good potential into the 21st century [281].

In spite of the fact that there is significant improvement in particulate filtration along with various sorbents and associated equipment over the past few years, there still seems to be a long way to go to achieve complete reliability of these systems, which have never been tested successfully in a real integrated gasification (IG)-based system environment. So far, a robust and completely reliable technology has not been developed especially for gas cleaning over 600°C to achieve higher efficiencies. As reported, most of these candle filters appears to have operated for a maximum period of 2700 hr at 400°C and little longer period of about 15,000 hr at lower temperature of 285°C in a coal-gasification environment. The failing of the candle filters within a short period of operation leads to an uneconomical plant availability factor [282]. For longer filter life, it is important to determine the consistency in manufactured qualities. A dynamic characterization method is recommended as a nondestructive evaluation technique [283]. However, researchers in the area should be aware that there are certain fundamental limitations to improve the intrinsic material properties of candle filters, and therefore alternate routes involving operating conditions, and process-related changes should be explored to develop more reliable and efficient gas-cleaning technology.

Pollution control strategies in developed countries continue to evolve in response to new and changing environmental regulations, new technologies, and even some new applications of proven technologies. Control of Hg emissions from coal-fired power plants is gaining increasing attention in several parts of the world. Some of the future changes expected in the environmental regulations include tighter emission regulations on SO_2, NO_x, $PM_{2.5}$, Hg control, and air toxics such as lead and cadmium. There is a continual refining and upgrading of product offerings for $DeNO_x$ Selective Catalytic Reduction (SCR), pulse-jet fabric filters, and rigid discharge electrode ESPs. To provide a more complete product line of air pollution control technologies, an additional type of lime spray dryer absorber for dry FGD, new wet electrostatic precipitator (WESP) technologies, and single tower *in situ* forced oxidation wet FGD (IFO-WFGD) systems have also come up [284]. It may be added that the use of nonthermal plasma, which is used in laboratory, might show its potential in near future.

In line with the above, future development in pulse-jet filtration will encompass the following areas:

- A more compact particulate collector;
- A more energy-efficient particulate collector;
- A device with high particulate-removal efficiency (99.99%) for all particle sizes including fine particles;

- Development of suitable sorbent to control gaseous emission;
- Development of new technique for simultaneous control of particulate and gaseous pollutants;
- Improved life of filter element.

Acknowledgements

The author and publisher acknowledge the help of various organizations, such as Hamon Research-Cottrell, Inc., USA; ADA – Environmental Solution, USA; Electric Power Research Institute, USA; Filtration Testing Equipment and Services GmbH; and Ankersmid Ltd., the Netherlands, for providing some useful graphic materials and other information incorporated in the paper. Various publishers, such as Elsevier B.V.; Springer Science and Business Media, Heidelberg; Sage Publication Ltd., UK; Melliand Textilberichte, Germany, are also duly acknowledged for their contributions and giving kind permission to reproduce figures and tables that originally appeared in their journals. The author wishes to put on record the extensive use of ScienceDirect (Elsevier B.V.), and consulting the website materials from the Department of Energy (DOE), USA; Environment Protection Agency (EPA), USA; Institute of Clean Air Companies (ICAC) USA; Alstom, USA; Babcock & Wilcox, USA; Siemens, USA; Donaldson, USA; Gore & Associates, USA; Evonik Fibres, GmbH; Menardi, USA; ASCO Numetics, France; PALAS, GmbH, and many others. The authors of various publications consulted for the preparation of this paper are also gratefully acknowledged.

References

[1] T. Godish, *Air Quality*, Lewis Publishers (A CRC Press Company), Boca Raton, FL, 2003.

[2] R. Trzupek, *Air Quality Compliance and Permitting Manual*, McGraw-Hill Professional, New York, USA, 2002.

[3] Air quality management online portal, US Environmental Protection Agency. Available at www.epa.gov/air/aqmportal/management/control_strategies.htm, accessed on April 24, 2009.

[4] N.P. Cheremisinoff, *Handbook of Air Pollution Prevention and Control*, Butterworth-Heinnemen (an imprint of Elsevier Indian reprint), New Delhi, India, 2002.

[5] C.E. Baukal, Jr., *Industrial Combustion Pollution and Control*, Marcel Dekker, New York, NY, 2004.

[6] J.D. McKenna, J.H. Turner, and J.P. McKenna, *Fine Particles (2.5 Microns) Emissions*, John Wiley and Sons, Hoboken, NJ, 2008.

[7] Fine particle ($PM_{2.5}$) designations, US Environmental Protection Agency. Available at http://www.epa.gov/pmdesignations/, accessed on June 10, 2009.

[8] W. Winkenwerder, Environmental exposure report on particulate matter (October 15, 2002). Available at http://www.gulflink.osd.mil/particulate_final/particulate_final_s05.htm – 26*k*, accessed on June 10, 2009.

[9] Council directive 1999/30/EC (22 April 1999) relating to limit values for sulphur dioxide, nitrogen dioxide and oxides of nitrogen, particulate matter and lead in ambient air, Official J. European Community. Available at http://europa.eu/legislation_summaries/environment/air_pollution/l28098_en.htm, accessed on June 29, 1999.

[10] Directive of the European parliament and of the council on ambient air quality and cleaner air for Europe, European Commission. Available at http://ec.europa.eu/environment/archives/air/cafe/pdf/cafe_dir_en.pdf, accessed on June 29, 2009.

[11] National ambient air quality standards, US Environmental Protection Agency. Available at http://www.epa.gov/air/criteria, accessed on February 11, 2008.

[12] H.S. Peavy, D.R. Rowe, and G. Tchobanoglous, *Environmental Engineering*, McGraw-Hill International Edition, New York, USA, 1985.

[13] T.K. Ray, *Air Pollution Control in Industries – Volume I, Theory, Selection and Design of Air Pollution Control Equipments*, Tech Books International, New Delhi, India, 2004.

[14] E. Riser-Roberts, *Remediation of Petroleum Contaminated Soils*, Lewis Publisher, Boca Raton, FL, USA, 1998.
[15] D.v. Velzen, H. Langenkamp, and G. Herb, Waste Manage. Res. 20 (2002), p. 556.
[16] K.R. Parker, J. Power Energy 211 (1997), p. 53.
[17] K.F. Maxey and M.J. Rosenau, *Preventive Medicine and Public Health*, 9th ed., Appleton-Century-Craft, New York, NY, 1965.
[18] L. Morawska, M.R. Moore, and Z.D. Ristovski, Health impacts of ultrafine particles, Department of the Environment and Heritage, Commonwealth of Australia 2004. Available at http://www.environment.gov.au/atmosphere/airquality/publications/health-impacts/pubs/health-impacts.pdf, accessed on April 24, 2009.
[19] V. Westergaard, Sanitary bag filter-SANICIP. [Brochure] 2003 [Cited 26/3/04]. Available at www.niro.com/ndk_website/NIRO/CMSResources,nsf/filenames/160303_SANICIP.pdf/$file/160303_SANICIP.pdf, accessed on June 15, 2009.
[20] M.N. Rao and H.V.N. Rao, *Air Pollution*, Tata McGraw-Hill Publishing, New Delhi, India, 1989.
[21] W.L. Heumann, *Industrial Air Pollution Control Systems*, McGraw-Hill Professional, New York, USA, 1997.
[22] L.K. Wang, C. Williford, and W.-Y. Chen, *Air pollution control engineering,* in *Handbook of Environmental Engineering*, Vol. 1, L.K. Wang, N.C. Pereira, and Y.-T. Hung, eds., Humana Press, New Jersey, 1st ed., 2004.
[23] R.W. McIlvaine, M. Carpenter, and S. Reinhold, Filtr. Separat. 24 (1987), p. 342.
[24] C.E. Baukal, Jr., *Industrial Combustion Pollution and Control*, Marcel Dekker, New York, NY, 2004.
[25] Mercury control with fabric filters from coal-fired boilers, The Institute of Clean Air Companies. Available at http://www.icac.com/files/public/ICAC_Hg_Control_with_FF_051606.pdf, accessed on January10, 2009.
[26] H. Yi, J. Hao, and L. Duan, Fuel 87 (2008), p. 2050.
[27] R. Wahl and E. Walz, *Method and apparatus for removing dust and gas pollutants from waste gases, particularly waste gases produced in the manufacture of optical waveguide performs*, US Patent 4765805, August 1988.
[28] G. Ruoff, *Method and device for the adsorption and chemisorptions respectively of gaseous components in a gas stream*, US Patent 5387406, February 1995.
[29] P. Chu, W. Downs, and J.B. Doyle, *Improved* SO_x, NO_x, *and particulate removal system*, US Patent 5540897, July 1996.
[30] R. Martinelli, D.W. Johnson, and R.B. Myers, *Flue gas conditioning for the removal of acid gases, air toxics and trace metals*, US Patent 5599508, February 1997.
[31] B.J. Lerner, *Method for minimizing environmental release of toxic compounds in the incineration of wastes*, US patent 5, 607, 654, March 4, 1997.
[32] S.C. Carlton, R.V. Huff, and C.F. Hammel, *System for removal of pollutants from a gas stream*, US Patent 7247279, July 2007.
[33] Ross & Associates Environmental Consulting, *Draft Report for Mercury Reduction Options*, Prepared for: US Environmental Protection Agency, September 1, 2000. Available at http://www.epa.gov/bns/mercury/Draft_Report_for_Mercury_Reduction_Options.pdf, accessed on September 05, 2009.
[34] N.L. de Freitas, J.A.S. Gonçalves, and M.D.M. Innocentini, J. Hazard. Mater. 136 (2006), p. 747.
[35] K.N. Chatterjee, A. Mukhopadhyay, and S.C. Jhalani, Indian J. Fibre Text. Res. 21 (1996), p. 194.
[36] I.C. Sharma, K.N. Chatterjee, and A. Mukhopadhyay, Indian J. Fibre Text. Res. 23 (1998), p. 38.
[37] L. Lillieblad, P. Wieslander, and J. Hokkinen, $PM_{2.5}$ and mercury emissions from a high ratio fabric filter after a pulverized coal fired boiler. Available at www.icac.com/files/public/MEGA03_200.pdf, accessed on January 10, 2009.
[38] N.P. Cheremisinoff, *Handbook of Air Pollution Prevention and Control*, Butterworth-Heinemann (An imprint of Elsevier), Burlington, MA, USA, 2002.
[39] S.K. Friendlander, Environment 31 (1989), p. 10, 36.
[40] S. Shanthakumar, D.N. Singh, and R.C. Phadke, Fuel 87 (2008), p. 3216.
[41] D. Eskinazi, J.E. Cichanowicz, and W.P. Linak, J. Air Pollut. Control Assoc. 39 (1988), p. 1131.

[42] M. Hupa, R. Backman, and S. Bostrom, J. Air Pollut. Control Assoc. 39 (1989), p. 1496.
[43] J.F. Skea and E.S. Rubin, J. Air Pollut. Control Assoc. 38 (1988), p. 1281.
[44] L. Zhang, Y. Zhuo, L. Chen, X. Xu, and C. Chen, Fuel Process. Technol. 89 (2008), p. 1033.
[45] A. Kolker, C.L. Senior, and J.C. Quick, Appl. Geochem. 21 (2006), p. 1821.
[46] Y. Wang, Y. Duan, and L. Yang, Fuel Process. Technol. 90 (2009), p. 643.
[47] J.E. Cichanowicz, P.M. Nasoney, and S.J. Davidson, J. Air Pollut. Control Assoc. 38 (1988), p. 1222.
[48] H.J. White, *Industrial Electrostatic Precipitation*, Addison-Wesley, 1963. Reading, MA, USA, pp. 128–135.
[49] A. Jaworek, A. Krupa, and T. Czech, J. Electrostat. 65 (2007), p. 133.
[50] N. Biege, H. Pretti, and W. Pirwitz, Emerging and competing technologies for particulate control in cement plants, 39th IEEE Cement Industry Technical Conference, 1997 IEEE-IAS. Available at http://ieeexplore.ieee.org/iel3/4694/13112/00599364.pdf?arnumber=599364, accessed on June 24, 2009.
[51] S.L. Francis, A. MacPherson, and M. Generation, $PM_{2.5}$ and mercury emissions from a high ratio fabric filter. Available at www.power.alstom.com/home/equipment systems/ecs/power/fabric_filter /technical_papers/_files/file_36682_42981.pdf, accessed on April 12, 2009.
[52] T.K. Ray, *Air Pollution Control in Industries – Volume II, Application of Air Pollution Control Equipments*, Tech Books International, New Delhi, India, 2004.
[53] W.T. Davis, A.J. Buonicore, and L. Theodore, *Air pollution control engineering*, in *Air Pollution Engineering Manual*, W.T. Davis, ed., John Wiley, New York, NY, 2000.
[54] Y. Wang, Y. Duan, and L. Yang, J. Fuel Chem. Technol. 36 (2008), p. 23.
[55] F. Scala and H.L. Clack, J. Hazard. Mater. 152 (2008), p. 616.
[56] J.D. Kilgroe, J. Hazard. Mater. 47 (1996), p. 163.
[57] A.T. Agarwal, Chem. Eng. 112(2) (2005), p. 42.
[58] D.S. Leith and M.W. First, J. Air Pollut. Control Assoc. 27 (1977), p. 754.
[59] D.S. Leith, M.W. First, and D.D. Gibson, Filtr. Separat. 15 (1978), p. 400.
[60] S. Moore, J. Rubak, and M. Jolin, Plant Eng. 50(11) (1996), p. 58.
[61] L. Morgan and M. Walters, Ceram. Ind. 149(7) (1999), p. 24.
[62] T.J. Vidmar, Plant Eng. 44(8) (1990), p. 70.
[63] M.G. Cora and Y. Hung, Environ. Qual. Manage. 11(4) (2002), p. 53.
[64] A.C. Caputo and P.M. Pelagagge, Environ. Prog. 19 (2000), p. 238.
[65] A. Ogawa, *Separation of Particles from Air and Gases*, CRC Press, Boca Raton, FL, 1984.
[66] R.W. McIlvaine, *Fabric Filter Market Rises, Replacing Precipitators, Environment Solutions*, November (1995), Environment Solutions, Texas, USA, p. 21.
[67] A.C. Caputo and P.M. Pelagagge, Environ. Manage. Health 10 (1999), p. 96.
[68] R.L.R. Salcedo, V.G. Chibante, and A.M. Fonseca, Powder Technol. 172 (2007), p. 89.
[69] K. Wark and C.F. Warner, *Air Pollution: Its Origin and Control*, 2nd ed., Harper and Row, New York, NY, 1981.
[70] M.A. Cuenca and E.J. Anthony, *Pressurized Fluidized Bed Combustion*, Blackie Academic & Professional, Glasgow,UK, 1995, pp. 211–254.
[71] W. Peukert, Filtr. Sep. 35 (1998), p. 461.
[72] Z.L. Ji, M.X. Shi, and F.X. Ding, Powder Technol. 139 (2004), p. 200.
[73] J.P.K. Seville, T.G. Chuah, and V. Sibanda, Adv. Powder Technol. 14 (2003), p. 657.
[74] D. Koch, K. Schulz, and J.P.K. Seville, *Regeneration of rigid ceramic filters*, in *Proceedings of the 2nd International Symposium on Gas Cleaning at High Temperatures*, R. Clift and J.P.K. Seville, eds., Blackie Academic & Professional, Glasgow, UK, 1993, pp. 244–265.
[75] V.H. Belba, W.T. Grubb, and R. Chang, J. Air Waste Manage. Assoc. 42 (1992), p. 209.
[76] A. Kavouras and G. Krammer, Powder Technol. 133 (2003), p. 134.
[77] A. Ergüdenler, W. Tang, and C.M.H. Brereton, Sep. Purif. Technol. 11 (1997), p. 1.
[78] K.N. Chatterjee, A. Mukhopadhyay, and S.C. Jhalani, Indian J. Fibre Textile Res. 22 (1997), p. 13.
[79] Y.-H. Cheung and C.-J. Tsai, Aerosol Sci. Technol. 29 (1998), p. 315.
[80] C.-J. Tsai, M.L. Tsai, and H.C. Lu, Sep. Sc. Technol. 35 (2000), p. 211.
[81] R.P. Donovan, *Fabric Filtration for Combustion Sources – Fundamentals and Basic Technology*, Marcel Dekker, New York and Basel, 1985.
[82] W. Humphries, M. Miceli, and J.J. Madden, Text. Res. J. 54 (1984), p. 237.

[83] F. Löffler and J Sievert, Filtr. Sep. 24 (1987), p. 110.
[84] F.J. Gutiêrrez Ortiz, B. Navarrete, and L. Cañadas, Chem. Eng. J. 127 (2007), p. 131.
[85] K. Everaert, J. Baeyens, and J. Degreve, J. Air Waste Manage. Assoc. 52 (2002), p. 1378.
[86] H.K. Choi, C. Lee, and H.K. Lee, Korean J. Chem. Eng. 24 (2007), p. 361.
[87] K.T. Hindy, Atmos. Environ. 20 (1986), p. 1517.
[88] O. Bjarnø and L. Linandau, *Method when cleaning a filter*, US Patent 6749665 B2, June 15, 2004.
[89] F. Löffler and R. Klingel, Filtr. Sep. 20 (1983), p. 205.
[90] E. Hardman, *Textile in filtration*, in *Handbook of Technical Textiles*, A.R. Horrocks and S.C. Anand, eds., Woodhead Publishing, The Textile Institute, CRC Press, UK, Boca Raton, FL, USA, 2000.
[91] W. Gregg, R. Vendetti, and R. Lindsay, *Low pressure pulse jet dust collector*, US Patent 5421845, June 6, 1995.
[92] Industrial filtration solutions, Donaldson Company. Available at www.donaldsontorit.com, accessed on September 08, 2009.
[93] G.D. Lanois and A. Wiktorsson, *Current status and future potential for high-ratio fabric filter technology applied to utility coal-fired boilers*, in *Proceedings of the First Conference on Fabric Filter Technology for Coal-Fired Power Plants,* Denver, CO, 1982, pp. 4125–4154.
[94] C.J. Bustard, K.M. Crushing, and R.L Chang, J. Air Waste Manage. Assoc. 42 (1992), p. 1240.
[95] W. Humphries and J.J. Madden, Filtr. Sep. 20 (1983), p. 40.
[96] H.-C. Lu and C.-J. Tsai, J. Environ. Eng. 129 (2003), p. 811.
[97] Fabric filter, ALSOM Power. Available at www.power.alstom.com/_looks/alstomV2/frontofficeScripts/index.php?languageId=EN&dir=/.../power/fabric_filter/, accessed on June 12, 2008.
[98] L.L. Lavely and A.W. Ferguson, *Power plant atmospheric emission control (Chapter 14)*, in *Power Plant Engineering*, L.F. Drbal, P.G. Boston, and K.L. Westra, eds., Kluwer Academic Publishers, Dordrecht, Netherlands, 1996, pp. 418–463.
[99] R. Sims, Dust collection design: incorporating safety, performance, and energy savings, Clyde Materials Handling, Incorporating MAC Equipment. Available at http://www.bulk-online.com/Forum/showthread.php?threadid=17921, accessed on September 08, 2009.
[100] T. Golesworthy, Filtr. Sep. 36 (5) (1999), p. 24.
[101] *Coated fabric cartridge filter medium for dust filtration*, Lohmann GmbH & Co., Lohmann Technologies UK Ltd., and Ravensworth Ltd., Filtr. Sep. 30 (1993), pp. 613–614.
[102] C.R. Smith, Filtr. Sep. 31 (1994), p. 179.
[103] *OAQPS Control Cost Manual*, Fifth Edition, Chapter 5, Office of Air Quality Planning and Standards, US EPA 453/B-96-001, Research Triangle Park, NC, December 1998.
[104] *Stationary Source Control Techniques Document for Fine Particulate Matter,* Office of Air Quality Planning and Standards, EPA-452/R-97-001, Research Triangle Park, NC, October 1998.
[105] Air pollution control fact sheet, US Environmental Protection Agency. Available at EPA-452/F-03-004, http://www.epa.gov/ttn/catc/dir1/ff-cartr.pdf, accessed on June 28, 2009.
[106] Industrial dust collection systems, Torit Products (a Subsidiary of Donaldson Company, Inc.). Available at www.torit.com/lproducts, accessed on February 14, 2009.
[107] J.H. Turner, J.D. McKenna, and J.C. Mycock, *Fabric Filters*, Chapter 5, Research Triangle Institute and US Environmental Protection Agency, December 1998. Available at www.p2pays.org/ref/10/09848.pdf, accessed on June 22, 2009.
[108] Technical Information, Menardi Filters, SC. Available at www.menardifilters.com, accessed on July 01, 2009.
[109] Control Technology Information – Fabric Filters, January 11, 1999, Institute of Clean Air Companies, USA. Available at www.icac.com, accessed on June 22, 2009.
[110] Pactecon® Envelope Filter Collectors, Sly Inc., US. Available at www.slyinc.com/pages/dryfilter/pactecon.html, accessed on June 30, 2009.
[111] Z. Ji, M. Shi, and F. Ding, Powder Technol. 139 (2004), p. 200.
[112] H. Sasatsu, N. Misawa, and M. Shimizu, Powder Technol. 118 (2001), p. 58.
[113] M.A. Alvin, Fuel Process Technol. 56 (1998), p. 143.
[114] A. Startin and G. Elliott, Filtr. Sep. 38(9) (2001), p. 38.

[115] K. Sutherland, *Filters and Filtration Handbook*, 5th ed., Butterworth-Heinmann, Burlington, MA, USA, 2008.
[116] J.F. Zievers, P.M. Eggerstedt, and P. Aguilar, Filtr. Sep. 27 (1990), p. 353.
[117] J.-H. Choi, Y.-G. Seo, and J.-W. Chung, Powder Technol. 114 (2001), p. 129.
[118] A. Dittler, M.V. Ferer, and P. Mathur, Powder Technol. 124 (2002), p. 55.
[119] D.H. Smith, V. Powell, and G. Ahmadi, Powder Technol. 94 (1997), p. 15.
[120] H. Kamiya, Y. Sekiya, and M. Horio, Powder Technol. 115 (2001), p. 139.
[121] Control of particulate matter emissions: APTI 413, Student Manual, Chapter 7, US Environmental Protection Agency. Available at http://www.epa.gov/apti/Materials/APTI%20413%20student/413%20Student%20Manual/SM_ch%207.pdf, accessed on May 20, 2009.
[122] Mikropul product bulletin, Mikropul USA. Available at www.mikropul.com/products/pdf_files/mikro_pulsaire.pdf, www.mikropul.com/products/media/poptop2.pdf, accessed on May 25, 2009.
[123] H.F. Johnson, *Fabric filter with gas inlet geometry and method*, US Patent 5846300, December 1998.
[124] G. Giusti and R.W. Duyckinck, *Gas inlet construction for fabric filter dust collections*, US Patent 4883510, November 1989.
[125] J.A. Cross, R. Helstroom, and R. Beck, Filtr. Sep. 32 (1995), p. 443.
[126] H.-C. Lu and C.-J. Tsai, Environ. Sci. Technol. 30 (1996), p. 3243.
[127] K.T. Hindy, J. Sievert, and F. Löffler, Environ. Int. 13 (1987), p. 175.
[128] P. Gang, Filtr. Sep. (Inter. Ed.) 9 (2009), p. 17.
[129] D. Purchas and K. Sutherland, *Handbook of Filter Media*, Elsevier Advanced Technology, Kidlington, UK, 2001.
[130] S.J. Miller and D.L. Landal, Pulse-jet baghouse performance improvement with flue gas conditioning, DOE Scientific and Technical Information, October 1992. Available at http://www.osti.gov/bridge/product.biblio.jsp?query_id=0&page=0&osti_id=10142658, accessed on October 04, 2009.
[131] E. Schmidt and T. Pilz, Filtr. Sep. 33 (1996), p. 409.
[132] K. Morris and R.W.K. Allen, Filtr. Sep. 33 (1996), p. 339.
[133] R.C. Carr, and W.C. Smith, J. Air Pollut. Control Assoc. 34 (1984), p. 281.
[134] S.K. Bandopadhyay, M.V. Ranga Rao, and S.N. Ghosh, *Cement science and technology: Energy trends*, in *Progress in Cement and Concrete – Modernization and Technology Upgradation in Cement Plant*, S.N. Ghosh and Kamal Kumar, eds., Vol. 5, Akedemia Book International, New Delhi, India, 1999.
[135] W.J. Klimczack and G. Applewhite, Chem. Eng. Prog. 93 (1997), p. 56.
[136] W. van de Harr and T. Oomen, Filtr. Sep. 38(10) (2001), p. 36.
[137] L.G. Felix, *Apparatus and method for improved pulse-jet cleaning of industrial filters*, US Patent 6605139 B2, August 12, 2003.
[138] Pulse valves for dust collector systems, Emerson Electric, ASCO Numatics, France. Available at www.ascojoucomatic.it/images/site/upload/_en/.../X003jagb.pdf, accessed on July 11, 2009.
[139] Simatek product bulletin, Simatek A/S, Hoeng. Available at http://www.simatek.dk/documents/00003.pdf, accessed on May 25, 2009.
[140] K.C. Schifftner, *Air Pollution Control Equipment Selection Guide*, CRC Press LLC, Boca Raton, FL, 2002.
[141] Goyen innovative environmental solutions, Tyco International Ltd. Available at http://www.goyen.com/__data/page/11833/CAS_INN_ENV_SOL_042009.pdf, accessed on September 05, 2009.
[142] W.J. Morris, Filtr. Sep. 21 (1984), p. 50.
[143] E. Rothwell, Filtr. Sep. 28 (1991), p. 341.
[144] W. Humphries and J.J. Madden, Filtr. Sep. 18 (1981), p. 503.
[145] G.-M. Klein, T. Schrooten, T. Neuhaus, R. Esser, F. Ott, and T. Daniel, *Enhanced energy efficiency solutions for industrial baghouse filters*, Presented at the International Conference and Exhibition for Filtration and Separation, FILTECH – 2009, Wiesbaden, Germany, October 13–15, 2009, pp. 144–151.
[146] L. Morgan and C. Farr, Recirculating air from dust collector, Plant Engineering, September 2001. Available at http://www.farrapc.com/literature/articles/recirculating-dust-collectors.pdf, accessed September 08, 2009.

[147] G.E.R. Lamb and P.A. Costanza, Filtr. Sep. 17 (1980), p. 319.
[148] Results of activated carbon injection for mercury control upstream of a COHPAC Fabric Filter, Hamon Research Cottrel, USA. Available at http://hamon-researchcottrell.com/COHPACTMandTOXECONTM.asp, accessed on May 19, 2009.
[149] A.F. Hollingshead, Dust collector design and Safety, Cement Industry Technical Conference Record, 2008 IEEE, May 18–22, 2008, IEEE, 2008. Available at ieeexplore.ieee.org/iel5/4539594/4539595/04539605.pdf?
[150] *ATEX DIRECTIVE 94/9/EC for Equipment intended for use in Potentially Explosive Atmospheres (ATEX),* European Commission, Enterprise and Industry, 3rd Edition, June 2009, ec.europa.eu/enterprise/atex/guide/index.htm.
[151] M. Althouse and N. Egbert, Dust collection design: incorporating safety, performance, and energy savings. Available at http://www.macequipment.com/pdf/Dust%20Collection%20Design.pdf, accessed on September 08, 2009.
[152] Occupational safety and health administrations, US. Department of Labor, Washington DC. Available at http://www.osha.gov/dsg/combustibledust/index.html, accessed on September 08, 2009.
[153] Dust collector systems and equipment for air cleaning, Emerson Electric Co., Asco Numatics, France. Available at www.ascojoucomatic.nl/images/site/upload/_en/.../X003agb.pdf, accessed on July 11, 2009.
[154] D.C. Draemel, *Relationship between fabric structure and filtration performance in dust filtration*, Project Number: EPA-21-ADJ-51, National Technical Information Service, US Department of Commerce, Alexandria, VA, USA, July 1973.
[155] Fabric filter material, lesson 4, Environmental Protection Agency, US. Available at http://yosemite.epa.gov/oaqps/eogtrain.nsf/b81bacb527b016d785256e4a004c0393/c4aba387f9325dca85256b6d004f7d96/$FILE/si412a_lesson4.pdf, accessed on October 09, 2009.
[156] T.C. Savage, R.C. Carlozzi, and O. Petzoldt, Enhanced aramid felt filter bags with new expanded PTFE membrane Technology. Available at www.gore.com/filtration/GORE-TEX@industrialbaghouse/article, accessed on January 18, 2007.
[157] Applications of ePTFE, W.L. Gore & Associates, US. Available at www.gore.com/en_xx/technology/timeline/applications_eptfe.html, accessed on February 20, 2009.
[158] Membrane Filters, Mueller sales US. Available at www.muellersales.com/pdf/sen_ptfe_polypro.pdf, accessed on February 20, 2009.
[159] Technical Capabilities: Expanded PTFE, W.L. Gore & Associates, US. Available at www.gore.com/en_xx/technology/creative/technical_capabilities.html, accessed on February 20, 2009.
[160] DuPont™Teflon® PTFE TE-3864, DuPont USA, http://www2.dupont.com/Teflon_Industrial/en_US/assets/downloads/k15759.pdf, accessed on July 1, 2009.
[161] P.A. Smith, *New developments in nonwovens for air and gas filtration,* International Seminar on Nonwovens, Bombay Textile Research Association, January 22–23, 1990.
[162] GORE® High Durability Filter Bags, W.L. Gore & Associates, US. Available at www.gore.com/en_xx/products/filtration/baghouse/filterbags/highdurability/hd_filter_bags.html, accessed on February 19, 2009.
[163] E. Schmaiz, M. Sauer-Kunze, and L. Bergmann, *Filtration*, in *Nonwoven Fabric*, W. Albrecht, H. Fuchs, and W. Kittelmann, eds., WILEY-VCH Verlag GmbH & Co. KgaA, Weinheim, Germany, 2003, pp. 557–570.
[164] J.R. Gabites, J. Abrahamson, and J.A. Winchester, Powder Technol. 187 (2008), p. 46.
[165] F. Löffler, H. Dietrich, and W. Flatt, *Dust Collection with Bag Filters and Envelope Filters*, Friedr. Vieweg & Sons, Braunschweig/Wiesbaden, Germany, 1988.
[166] Y. Otani, H. Emi, and J. Mori, J. Aerosol Sci. 22 (1991), p. 793.
[167] N. Plaks, J. Electrostat. 20 (1988), p. 247.
[168] T.C. Dickenson (ed.), *Filters and Filtration Handbook*, 4th ed., Elsevier Advanced Technology, Oxford, UK, 1997.
[169] M. Neate, Laboratory Report – Coal Fired Power Station, Advancetex Laboratory, Australia, September 22, 2009. Available at http://www.advancetex.net/downloads/PowerStationLaboratoryReport.pdf
[170] A. Wimmer, Filtr. Sep. 36(3) (1999), p. 26.
[171] R. O'Connor, Filtr. Sep. 34 (1997), p. 834.

[172] M. Schobesberger, R. Hemmelmayr, and W. Themmel, Evonik Fibres Gmbh. Available at http://www.p84.com/downloads/publications/, accessed on September 05, 2009.
[173] J. Johnson and J. Handy, *Breathing new life into dust collection filters*, Ceram. Ind. February (2003), pp. 39–41.
[174] O. Medvedyev and Y. Tsybulya, Filtr. Sep. 42(1) (2005), p. 34.
[175] K.M. Cushing, W.T. Grubb, and B.V. Corina, Long-Term COHPAC Baghouse Performance at Alabama Power Company's E. C. Gaston Units 2 & 3. Available at http://secure.awma.org/presentations/Mega08/Papers/a16_1.pdf, accessed on September 05, 2009.
[176] K.N. Chatterjee, A. Mukhopadhyay, and S.C. Jhalani, Indian J. Fibre Text. Res. 22 (1997), p. 13.
[177] H.L. Huang, Y.C. Huang, and H.M. Wang, *Aerosol filtration efficiency of teflon fibrous filters*, Abst. Eur. Aerosol Conf. S973–S974, 2004.
[178] K.N. Chatterjee, A. Mukhopadhyay, and S.C. Jhalani, Indian J. Fibre Text. Res. 22 (1997), p. 21.
[179] *Applications: Filtration*, Nonwoven Markets, International Fact book and Directory, 1996, pp. 135–140.
[180] K.N. Chatterjee, A. Mukhopadhyay, and S.C. Jhalani, Indian J. Fibre Text. Res. 21 (1996), p. 251.
[181] S.Y. Yeo, O.S. Kim, and D.Y. Lim, J. Mater. Sci. 40 (2005), p. 5393.
[182] Dura-Life^{TM} filter bag technology, Donaldson Europe B.V.B.A. Available at http://www.emea.donaldson.com/it/industrialair/support/datalibrary/050506.pdf, accessed on September 08, 2009.
[183] I. Holme, P.V.K. Lee, and K. Mehrotra, J. Text. Inst. 73 (1982), p. 25.
[184] C.A. Lawrence and P. Liu, Chem. Eng. Technol. 29 (2006), p. 957.
[185] C.T. Martin, *Fine Filtration Fabric Options Designed for Better Dust Control and to Meet $PM_{2.5}$ Standards*, 1999 IEEE-IAS/PCA Cement Industry Technical Conference, April 11–15, 1999, Roanoke, Virginia, IEEE Catalog Number 99CH36335, pp. 385–393.
[186] J.W. Griffin, Donaldson Company, *Laminate and pulse jet filter bag*, US Patent 6517919, February 11, 2003.
[187] K.J. Fritsky, Long-term performance of GORE-TEX membrane filter bags at a municipal solid waste combustion facility, W.L. Gore & Associates. Available at http://www.gore.com/MungoBlobs/article_filter_municipal_solid_waste,6.pdf, accessed on September 05, 2009.
[188] S. Atkinson, *'Embedded membrane' creates robust filtration media*, Membr. Technol. No. 11 (2003), p. 8.
[189] Adjusting pulse system for optimum cleaning, W.L. Gore & Associates, US. Available at www.gore.com, accessed on February 21, 2009.
[190] Donaldson Filter Components Ltd., Filtr. Sep. 41(9) (2004), p. 14.
[191] Y.M. Jo, R. Huchison, and J.A. Raper, J. Waste Manage. Res. 14 (1996), p. 281.
[192] B.G. Miller, S.F. Miller, and R.T. Wincek, *A Demonstration of Fine Particulate and Mercury Removal in a Coal-Fired Industrial Boiler Using Ceramic Membrane Filters and Conventional Fabric Filters*, Presented at The EPRI-DOE-EPA Combined Utility Air Pollutant Symposium, The Mega Symposium, Atlanta, Georgia, August 16–20, 1999.
[193] B.G. Miller, R.T. Wincek, and D.C. Glick, *Ceramic Membrane Filters for Fine Particulate Removal in Coal-Fired Industrial Boilers,* Presented at the 23rd International Technical Conference on Coal Utilization & Fuel Systems, Clearwater, Florida, March 9–13, 1998, pp. 807–818.
[194] B.G. Miller, R.T. Wincek, and D.C. Glick, *A Comparison between Ceramic Membrane Filters and Conventional Fabric Filters for Fine Particulate Removal from a Coal-Fired Industrial Boiler*, Presented at the Fifteenth International Pittsburgh Coal Conference, Pittsburgh, PA, September 14–18, 1998.
[195] F. Goosens, J. Ind. Text. 22 (1993), p. 279.
[196] S. Gibbons, Filtr. Sep. 39(7) (2002), p. 20.
[197] R Lydon, Filtr. Sep. 41(9) (2004), p. 20.
[198] New cost-effective coatings extends filter life and efficiency, Ravensworth Ltd., UK. Available at http://wwt-magazine.info/news/news_story.asp?id=8929&channel=0, accessed on August 18, 2009.

[199] Ravlex increases filter-bag service life, Ravensworth Ltd., UK, Membrane Technology, 2004, No. 1 (2004) p. 4.
[200] V.K. Kothari, A. Mukhopadhyay, and S.N. Pandey, Melliand Textil. 74 (1993), p. 386.
[201] T. Ciach and L. Gradon, J. Aerosol Sci. 29 (1998), Supplementary Issue, p. S935.
[202] S. Calle', P. Contal, and D. Thomas, Powder Technol. 128 (2002), p. 213.
[203] J.H. Lin, C.W. Lou, and C.H. Lei, Composites Part A 37 (2006), p. 31.
[204] J.P. Fagan, *Felt-like layered composite of PTFE and glass paper*, US Patent 4,324,574, April 1982.
[205] F. Dotti, A. Varesano, and A. Montarsolo, J. Ind. Text. 37 (2007), p. 151.
[206] K. Kosminder and J.S. Scott, Filtr. Sep. 39(6) (2002), p. 20.
[207] S.Y. Yeo, D.Y. Lim, and S.W. Byun, J. Mater. Sci. 42 (2007), p. 8040.
[208] H.Y. Chung, J.R.B. Hall, and M.A. Gogins, *Bag house filter with fine fiber and spun bonded media*, US Patent 7318852, January 2008.
[209] W. Peukert and C. Wadenpohl, Powder Technol. 118 (2001), p. 136.
[210] S. Hajek and W. Peukert, *Progress of Hot Gas Cleaning in the Process Industry*, World Congress Particle Technology III, Brighton, UK, 1998.
[211] S. Zhu, R.H. Pelton, and K. Collver, Chem. Eng. Sci. 50 (1995), p. 3557.
[212] K. Morden, Int. Sugar J. 96(1143) (1994), p. 48.
[213] A.C. Handermann, *Basofil Filter Media – Efficiency Studies and an Asphalt Plant Baghouse Field Trial*, BASF Corporation, Fiber Products Division, Enka, NC, 1995.
[214] R. Sassa and R. Winkelmayer, *Static dissipative nonwoven textile material*, US Patent 5,213,882, May 1993.
[215] G.E.R. Lamb, P.A. Costanza, and D.J. O'Meara, Text. Res. J. 48 (1978), p. 566.
[216] K. Iinoya and K. Makino, J. Aerosol Sci. 5 (1974), p. 357.
[217] B.A. Kwetkus, J. Electrostat. 40–41 (1997), p. 657.
[218] D.A. Lundgren and K.T. Whitby, Ind. Eng. Chem. Process Des. Dev. 4 (1965), p. 345.
[219] H.J. Luckner, L. Gradoñ, and A. Podgórski, Inynieria Chemiczna i Procesowa 19 (1998), p. 891.
[220] A. Podgórski, H.J. Luckner, and L.Gradon', Inzynieria Chemiczna i Procesowa 19 (1998), p. 865.
[221] J.-K. Lee, S.-Ch. Kim, and J.-H. Shin, Aerosol Sci. Technol. 35 (2001), p. 785.
[222] N. Plaks, J. Electrostat. 20 (1988), p. 267.
[223] S. Kim, C. Sioutas, and M. Chang, Aerosol Sci. Technol. 32 (2000), p. 197.
[224] J. Mermelstein, S. Kim, and C. Sioutas, Aerosol Sci. Technol. 36 (2002), p. 62.
[225] R.F. Henry, W.F. Podolski, and S.C. Saxena, IEEE T. Ind. Appl. 21 (1985), p. 939.
[226] R.P. Donovan, L.S. Hovis, and G.H. Ramsey, Aerosol Sci. Technol. 1 (1982), p. 385.
[227] W. Humphries, C. Jones, and G. Miles, *Electrostatic Enhancement of a Fabric Filter Baghouse, Proceeding of the Second International Conference on Electrostatic Precipitation*, Institute of Electrostatics, Japan, Kyoto, 1984, p. 471.
[228] R.A. Fjeld and T.M. Owens, IEEE T. Ind. Appl. 24 (1988), p. 725.
[229] E.R. Frederick, J. Air Pollut. Control Assoc. 30 (1980), p. 426.
[230] O. Lastow and M. Bohgard, J. Aerosol Sci. 23(supplement 1) (1992), p. S105.
[231] S.-J. Yoa, Y.-S. Cho, and Y.-S. Choi, Korean J. Chem. Eng. 18 (2001), p. 539.
[232] H.-K. Choi, S.-J. Park, and J.-H. Lint, Korean J. Chem. Eng. 19 (2002), p. 342.
[233] R. Chang, *Compact hybrid particulate collector*, US Patent 5 158 580, October 17, 1992.
[234] R. Chang, *Compact hybrid particulate collector*, US Patent 5,024,681, January 18, 1991.
[235] W.A. Harrison, Fuel Energy Abstr. 38(3) (1997), p. 185.
[236] R.L. Miller and W.J. Morris, *Effective use of COHPAC technology as an effective multi-pollutant control technology*, USEPADOE – EPRI Combined Power Plant Air Pollution Control Symposium: The Mega Symposium, Chicago, IL, 20–23 August, 2001.
[237] R. Miller, R. Chang, and C.J. Bustard, *Effective use of both COHPAC™ and TOXECON™ technologies as the technology of the future for particulate and mercury control on Coal-fired boilers*, in Paper Presented at the 2003 International Power-Gen Conference, Las Vegas, NV, 2003. Available at http://www.hamon-researchcottrell.com/HRCTechnicalLibrary/MercuryControlonCoalFiredBoilers-COHPACTOXECON.pdf, accessed on June 15, 2009.
[238] R.E. Snyder and D.M. Novogoratz, Fabric filter size impacts on mercury control using activated carbon injection, Mega Symposium, Combined Power Plant Air Pollutant Control, Baltimore,

MA. Available at August 28–31, 2006, http://www.babcock.com/library/pdf/br-1786.pdf, accessed on January 23, 2009.
[239] S.J. Miller, G.L. Schelkoph, and G.E. Dunham, *Advanced hybrid particulate collector, A new concept for air toxics and fine-particle control*, US Department of Energy, National Energy Technology Laboratory (NETL). Available at www.netl.doe.gov/publications/proceedings/97/97ps/ps_pdf/PS2B-8.PDF, accessed on June 20, 2009.
[240] R. Gebert, C. Rinschler, and C. Polizzi, *A new filter system, combining a fabric filter and electrostatic precipitator for effective pollution control behind cement kilns*, Cement Industry Technical Conference, 2003. Conference Record IEEE-IAS/PCA 2003, 4–9 May 2003, pp. 285–294. Available at http://ieeexplore.ieee.org/stamp/stamp.jsp?arnumber=01204729.
[241] R. Gebert, C. Rinschler, and D. Davis, Commercialization of the advanced hybrid™ filter technology, US Department of Energy, National Energy Technology Laboratory (NETL). Available at http://www.netl.doe.gov/publications/proceedings/02/air_q3/Gerbert.pdf, accessed on June 20, 2009.
[242] Advanced hybrid particulate collector – pilot-scale testing, Technical Progress Report, W. Aljoe (Performance Monitor), US Department of Energy, National Energy Technology Laboratory, Pittsburgh, Cooperative Agreement No. DE-FC26-98FT40321, March 2003.
[243] A. Scheuch, *Dust filter with filter sleeve, emission electrode and collecting electrode*, USP 6869467, March 2005.
[244] H.V. Krigmont, L.J. Muzio, and R.A. Smith, *Multi-stage collector (MSC™) proof-of-concept pilot design and evaluation*, in Paper presented at ICESP IX – 9th International Conference on Electrostatic Precipitation, Pretoria, South Africa, 17–21 May, 2004.
[245] H.V. Krigmont, L.J. Muzio, and R.A. Smith, Multi-stage collector (MSC™) proof-of-concept pilot design and evaluation, Allied Environmental Technologies, Inc., CA. Available at www.alentecinc.com/papers/MSC/Krigmont_MSC-Pilot.pdf, accessed on July 02, 2009.
[246] H.V. Krigmont, *Multi-stage particulate matter collector*, US Patents 6,524,369 (February 2003) and US Patents 6,932,857 (August 2005).
[247] C.J. Bustard, M. Durham, and C. Lindsey, Results of activated carbon injection for mercury control upstream of a COHPAC fabric, The Institute of Clean Air Companies, Washington, DC. Available at http://www.icac.com/files/public/MEGA03_82_Hg.pdf, accessed on July 02, 2009.
[248] H.C. Hsi, M.J. Rood, and M. Rostam-Abadi, J. Environ. Eng. 128 (2002), p. 1080.
[249] S.J. Lee, Y.C. Seo, and J.S. Jurng, Atmos. Environ. 38 (2004), p. 4887.
[250] N.D. Hutson, B.C. Attwood, and K.G. Scheckel, Environ. Sci. Technol. 41 (2007), p. 1747.
[251] H.K. Choi, S.H. Lee, and S.S. Kim, Fuel Process. Technol. 90 (2009), p. 107.
[252] H. Yang, Z. Xu, and M. Fan, J. Hazard. Mater. 146 (2007), p. 1.
[253] D. Muggli, M. Durham, A. and O'Palko, Toxecon II™ and high-temperature reagents or sorbents for low-cost mercury removal. Available at http://www.icac.com/files/public/POWER-GEN_2005_Muggli.pdf, accessed on September 5, 2009.
[254] D.W. Agar, Chem. Eng. Sci. 54 (1999), p. 1299.
[255] F.M. Dautzemberg and M. Mukherjee, Chem. Eng. Sci. 56 (2001), p. 251.
[256] Y. Matatov-Meytal and M. Sheintuch, Appl. Catal. A-Gen. 231 (2002), p. 1.
[257] D. Fino, N. Russo, and G. Saracco, Chem. Eng. Sci. 59 (2004), p. 5329.
[258] O.H. Park, C.S. Kim, and H.H. Cho, Korean J. Chem. Eng. 23 (2006), p. 194.
[259] Catalytic filter system, ALSOM Power. Available at www.environment.power.alstom.com, accessed on January 20, 2008.
[260] G. Saracco and V. Specchia, *Structured Catalysts and Reactors*, A. Cybulsky and J.A. Moulijn, eds., Marcel Dekker, New York, NY, 1996, pp. 417–434.
[261] S.R. Ness, G.E. Dunham, and G.F. Weber, Environ. Progress 14 (1995), p. 69.
[262] G. Saracco and L. Montanaro, Ind. Eng. Chem. Res. 34 (1995), p. 1471.
[263] G. Saracco, S. Specchia, and V. Specchia, Chem. Eng. Sci. 51 (1996), p. 5289.
[264] G. Saracco and V. Specchia, Chem. Eng. Sci. 55 (2000), p. 897.
[265] J.H. Choi, S.-K. Kim, and Y.-C. Bak, Korean J. Chem. Eng. 18 (2001), p. 719.
[266] M. Hackel, G. Schaub, and M. Nacken, *Kinetics of reduction and oxidation reactions for application in catalytic gas-particle filters, in advanced gas cleaning technology*, in C. Kanaoka, H. Makino, and H. Kamiya, eds., *Proceedings of the 6th International Symposium on Gas Cleaning at High Temperatures*, October 20–22, 2005, Osaka, Japan, 2005, p. 37.

[267] M. Nacken, S. Heidenreich, and M. Hackel, Appl. Catal. B-Environ. 70 (2007), p. 370.
[268] J.J. Spivey, Appl. Catal. A-Gen. 105 (1993), p. N20.
[269] J.L. Bonte, K.J. Fritsky, and M.A. Plinke, Waste Manage. 22 (2002), p. 421.
[270] H.H. Glaze, J.W. Kang, and D.H. Chapin, Ind. Eng. Chem. Res. 28 (1989), p. 1580.
[271] G.T. Lee, J.H. Park, and B.H. An, J. Korean Soc. Environ. Eng. 20 (1998), p. 1599.
[272] Y. Ku, C.M. Ma, and Y.S. Shen, Appl. Catal. B-Environ. 34 (2001), p. 181.
[273] S. Heidenreich, M. Nacken, and M. Hackel, Powder Technol. 180 (2008), p. 86.
[274] H.V. Krigmont and Y. Akishev, *Multi-stage collector for multi-pollutant control,* US Patent Publication No. US 2008/0092736 A, April 2008.
[275] N. Harada, T. Moriya, and T. Matsuyam, J. Electrostat. 65 (2007), p. 37.
[276] N. Harada, T. Matsuyama, and H. Yamamoto, J. Electrostat. 65 (2007), p. 43.
[277] S. Masuda, S. Hosokawa, and X. Tu, J. Electrostat. 34 (1995), p. 415.
[278] R. Hackam and H. Akiyama, IEEE Trans., Dielectrics Elect. Insul. 7 (2000), p. 654.
[279] J. Benitez, *Process Engineering and Design for Air Pollution Control*, Prentice Hall, Englewood Cliffs, NJ, USA, 1993, p. 331 and p. 414.
[280] G. Elliott and A. Startin, Filtr. Sep. 34 (1997), p. 331.
[281] L. Bergmann, Filtr. Sep. 34 (1997), p. 137.
[282] S.D. Sharma, M. Dolan, and D. Park, Powder Technol. 180 (2008), p. 115.
[283] S.-E. Chen, Y. Nishihama, and P. Yue, Fuel 87 (2008), p. 2807.
[284] Current US air pollution control technologies & future trends, Hamon Research-Cottrell, Hamon USA. Available at http://hamon-researchcottrell.com/COHPACTMandTOXECONTM.asp, accessed on June 24, 2009.
[285] M.L. Croom, *Filter Dust Collectors: Design and Application*, McGraw-Hill, New York, USA, 1995.
[286] J.H. Turner, J.D. McKenna, and J.C. Mycock, Chapter 1: Baghouses and filters, Section 6: particulate matter controls, United States Environmental Protection Agency, 1998. Available at http://www.epa.gov/ttn/catc/dir1/cs6ch1.pdf, accessed on August 05, 2009.

Appendix

Calculations of filtering area and bag size – industrial practice

For the calculation of bag size, it is necessary to know the application type and various operational parameters such as the volumetric flow rate of the air stream, the operating temperature, concentration of dust particles and particle size, air-to-cloth ratio (ACR), etc. Based on the above, the following steps can be taken to find the bag size:

Step 1: *Conversion of given flow rate to the normal flow rate and deriving air-to-cloth ratio*

The actual air quantity can be converted to normal conditions based on the operating temperature and pressure. Q_{normal} (Nm^3/min) is the gas quantity at normal temperature and pressure conditions. Normal temperature (usually 0–20°C) and pressure (usually 1 atmospheric pressure) differs from place to place. Due to an increase in temperature, the gas volume increases, this obviously influences the gas characteristics. Hence, in practical terms, one has to convert the Q_{actual} to Q_{normal} following ideal gas equation. Assuming normal condition as 0°and 1 atmospheric pressure, Q_{actual} will be as follows:

$$Q_{\text{normal}} = Q_{\text{actual}} \times (273 \times P_{\text{act}}) / (273 +^{\circ}\text{C operating temperature}),$$

where P_{act} is the actual level of atmospheric pressure.

It may be noted that, in contrast to normal temperature and pressure (NTP), standard temperature and pressure (STP) refers to the condition at 20°C and 1 atmospheric pressure.

Air-to-cloth ratio is similar to filtering velocity (more specifically superficial filtration velocity, since air passes through an open area only).

Table AI. Correction factor for application.

S. No	Application	Correction factor
1	Nuisance venting: transfer points of conveyors, packing and filling stations	1.0
2	Product collection: driers, classifiers, grinders	0.9
3	Process gas exhaust: kilns, furnaces, spray driers, reactors	0.8

$$ACR = Q_{\text{actual}}\ (\text{m}^3/\text{s})\ /A\ (\text{m}^2),$$

where A = area of filter fabric in m^2.

Similarly, air-to-cloth ratio based on normal condition can be derived from

$$ACR = Q_{\text{normal}}\ (Nm^3/s)/A\ (m^2).$$

Clearly, the air-to-cloth ratio is one of the key baghouse design parameters. Several authors have a separate design procedure to establish suitable values for air-to-cloth ratios. Each is based on applying a number of factors that take into account the gas and dust properties when estimating an initial air-to-cloth ratio.

Step 2: *Derivation of actual filtration velocity* (V_f)

Depending on the application, one has to decide nominal/base filtering velocity (V_n) from the standard guideline [13], which can be converted into desired actual value (V_f) by using several correction factors. In the practical situation, it is necessary to deicide the correction factor (C_B) for the type of application situation: oily, moist, or agglomerative dust, product collection, and nuisance collection (Table AI).

In the case of particle size, higher velocity profile is opted for, for large-size particles and vice versa. The correction factor (C_P) for different range of particle size is given in Table AII.

Apart from the above-said factors, one has to consider the correction factors for temperature (C_T) and inlet dust concentration (C_D). Both these correction factors can be obtained from the chart given in Figure AI. For the assessment of correction factors for inlet dust

Table AII. Correction factor for particle size.

S. No.	Particle size (micron)	Correction factor
1	>100	1.2
2	50–100	1.1
3	10–50	1.0
4	5–10	0.9
5	2–5	0.8
6	<2	0.7

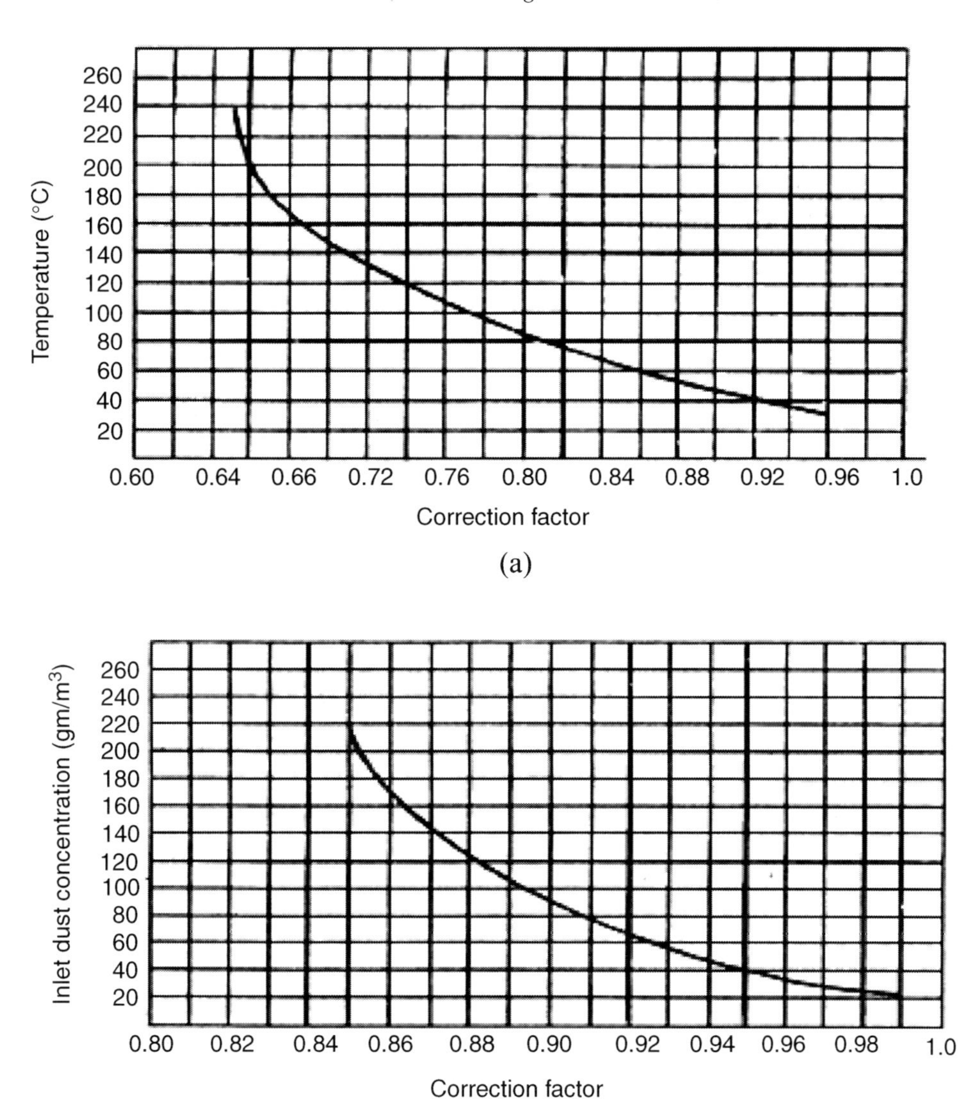

Figure AI. Correction factor. (a) For temperature; (b) For dust concentration.

concentration, the later parameter is normalized using the following formula:

$$C_{\text{normal}} = \frac{Q_{\text{actual}} \times C_{\text{actual}}}{Q_{\text{normal}}}.$$

Finally, the actual ACR can be calculated by multiplying all the correction factors with the velocity profile given by Croom [285]

$$V_f = V_n \times C_B \times C_P \times C_T \times C_D.$$

Löffler et al. [165] determined the effective air-to-cloth ratio, V_{fe} (m^3/m^2/min or m/min), from the following equation:

$$V_{fe} = V_n \times C_B \times C_P \times C_T \times C_D \times C_A \times C_F \times C_I \times C_H.$$

Like Croom [285], factors are applied for application type (C_B), particle size (C_P), gas temperature (C_T), and dust load (C_D). Additional factors are as follows:

- Filter system factor (C_A), distinguishing between single bag or group pulsing and online or offline cleaning.
- Bulk flow behavior (C_F), for dusts with bulk densities less than 600 kg/m^3.
- Flue gas flow (C_I), for online cleaning and upward gas flow past the bags.
- Tropical climate factor (C_H), applicable for food production and dusts with hygroscopic tendencies.

Yet, in another approach, Turner et al. [286] published a design procedure for the United States Environmental Protection Agency (USEPA) in 1998, with a specific method for pulse-jet baghouse. Again, factors for application type (B), gas temperature in degree Fahrenheit (T), dust loading in gr/ft^3 (D_T), and particle size in μm (P) are applied. The USEPA guidelines use the material factor, A_T, instead of a base air-to-cloth ratio (V_B), as used by Croom [285] and Löffler et al. [165], to distinguish between different dusts. Once the five factors, i.e. A_T, B, T, D_T, and P are determined, the recommended air-to-cloth ratio, V_{fe}, is calculated using the following equation:

$$V_{fe} = 2.878\, A_T B T^{-0.2335} D_T^{-0.06021} (0.7471 + 0.0853 ln P).$$

Step 3. Calculation of the total required filtering area

The total filteration area required for the given process condition can be calculated from the given volume flow rate (process requirement) (Q_{actual}) and the filtering velocity (V_f or V_{fe}).

$$\text{Total filtration area} = Q_{\text{actual}} / V_f.$$

Step 4. Calculation of bag dimension and number of bags

Considering the most common shape of the filter element (cylindrical bag shape), the available filtering area for each bag will be

$$Area = \pi D L + \pi \frac{D^2}{4},$$

Table AIII. Recommended maximum elutriation velocities.

Author	*Ve* (m/s)
Croom [285]	1.27
Agarwal [57]	1.52
Moore, Rubak, and Jolin [60]	1.83

where

D = diameter of the bag,
L = length of the bag.

The length of the bag is selected based on the effectiveness of cleaning on the full length of the bag. In determining the diameter/cross-sectional area of bag, one may check the elutriation velocity (V_e), particularly for a bottom entry-type filter unit. The upper limits of the elutriation velocity with their sources are given in Table AIII. The parameter can be determined by dividing the gas flow Q_{in} (m^3/min) by the cross-sectional area of the vessel (A_{BH} in m^2)– the cross-sectional area of the bag (A_{Bags} in m^2).

$$V_e = \frac{Q_{in}}{A_{BH} - A_{\mathrm{Bags}}}.$$

The basis for the elutriation velocities is a single-particle terminal velocity in the baghouse [57]. However, the terminal velocity of individual particles is strongly dependent on particle size. It is unclear which size should represent the powder falling fron the bags. This is expected to be in clusters of particles formed from the breakup of cake sections.

After calculating the filtering area available for the single bag structure, one can determine the number of bags required:

Total number of bags = Total filtering area available/Area of single bag.

If the calculated number of bags comes in a fraction, based on the nearest whole number, total filtering area and subsequent actual ACR can be calculated. Depending on the plant capacity, total gas volume varies widely; therefore the number of bags will also be different for similar applications.